"We forget that
the water cycle
and the life cycle
are one."
– Jacques Cousteau

It is time to remember.

Human impact

"Homo sapiens became prominent, maybe 200,000 years ago. Out of Africa they spread, maybe 60,000 years ago. And they began to have an impact. They wiped out the megafauna wherever they went, animals of maybe 100 pounds or more. Then, kaboom, kaboom, kaboom. Neolithic agriculture. Villages and towns and technology. We hit the world's fauna and flora with a sucker punch."

– E.O. Wilson, founder of the inquiry of sociobiology, and co-author of *The Theory of Island Biogeography*, and Pulitzer-prize winning co-author of *The Ants*

The Unraveling

Losing ecosystems

"Scientists from around the world warn that the planet is at risk of losing more than a million species in the coming decades, if swift action isn't taken to protect more of the natural world, stop exploitation of species, address climate change, reduce pollution, and stop the spread of alien invasive species.

All food and most medicines come directly from plants and animals. Species also form the building blocks of ecosystems, which purify air and water, pollinate crops, cycle nutrients, moderate climate, and more [ed.: build soil, produce oxygen, absorb greenhouse gasses, provide food and homes to wildlife, etc.]

In the United States, 40% of the nation's animals and 34% of the plants are threatened with extinction.

The International Union for Conservation of Nature found that 28% of plants and animals around the globe are threatened with extinction.

The IUCN Red List identifies 42,108 species as threatened, out of 150,388 species for which there is enough information to determine a conservation status.

Every lost species threatens to unravel ecosystems, and in the process reduce the services they provide."
– Center for Biological Diversity

With what has been – and is – happening to Earth and Nature, it is time for humanity to have a serious conversation, and to take immediate global actions to protect the environment and wildlife.

The agent is us

"Half of the world's few hundred primate species are threatened or endangered with extinction; and the problem, we know, is not at all limited to primates. With the obvious exception of domesticated animals, every major group of large mammals on the planet is declining in numbers. Both you and I know, in addition, that in virtually every case the agent behind this powerful global trend in extinctions is the same: us.'"

– Dale Peterson, *The Moral Lives of Animals*

The Sixth Great Extinction

The Holocene extinction

"We are currently experiencing the Sixth Great Extinction. Also known as the Holocene extinction, the Sixth Great Extinction is unique in its origins, because it is driven almost exclusively by human actions.

Based on a system that ranks extinction events based on the number of species lost, and the duration of the event, the current extinction event could likely be the worst in history.

It is likely that we will lose half of all plants, animals, and birds on our planet by the year 2100.

By not recognizing the importance of biodiversity – in addition to assuring the demise of most other species – we may be assuring the demise of our own human species as well.

We need biodiversity.

Our survival as a species ultimately rests on biodiversity. Each species plays a role in the global ecosystem, and we may not know until it is too late whether a given species was vital to our existence, but we can safely assume that we won't survive without a diversity of other species.

It is essential, now more than ever, that global action be taken on the level of government, organization, corporation, and individual to slow the rate of biodiversity loss."

– *The Sixth Great Extinction*; VoicesForBiodiversity.org

How far will the ecocide go? Is it leading to omnicide? That is, the end of humanity? What humans do from now on will decide.

We must act now

"According to multiple studies, Earth has entered its sixth large-scale extinction. Scientists from universities in the U.S. and Mexico found that species are disappearing at a rate 114 times faster than normal. The last time extinction rates were this high was when dinosaurs were wiped out more than 65 million years ago.

The International Union for Conservation of Nature keeps track of all known threatened species on their 'Red List,' which is backed by governments, scientists, and conservationists. According to them, almost one-third of all known species of plants and animals are at risk. The primary culprit is habitat loss from human-caused deforestation and development, but the authors of the new *Science Advances* study also say climate change is making the problem worse.

This mass extinction has major implications. Not only will it greatly impact ecosystems, as species that form every link in the food chain – from pollinators to predators – go extinct, but also humanity's ability to survive.

We depend on other species for everything from water purification to crop pollination to keeping insects in check. Without biodiversity, we will be in serious jeopardy of going extinct ourselves.

Each group of researchers has its own take on the crisis, and addresses the problem from different angles, but all come to the same conclusion: We must act now to stop the cascade of extinctions."

– Jennifer Collins, *We're Now Entering the Earth's Sixth Great Extinction*; EarthJustice.org

HUMANITY THE FINAL CENTURY

GLOBAL ECOCIDE AND THE HOLOCENE EXTINCTION

DANIEL JOHN CAREY

Author of
Dream Another Dream
and **Dream Your World**

Humanity:
The Final Century
Global Ecocide and the Holocene Extinction

Disclaimer:

Cover image by: Ricardo Honesto
Layout by: Noel

ISBN: 978-1-884702-29-7
Library of Congress Control Number: 2023952536
First Edition: 2024

Published by:
Oakonic, POB 1272, Santa Monica, CA 90406-1272, USA

For the love of Nature.

"Look deep into Nature,
and then you will understand everything better."
– Albert Einstein

Books by Daniel John Carey

Dream Your World

Dream Another Dream

Humanity: The Final Century

Plant-based Regenerative Nutrition

Screenwriting Tribe

Table of Contents

Part of ourselves

"For every species that is lost we have lost a part of ourselves. We become less alive when we cannot feel the tragedy that is unfolding before us.

In facing the reality of unconscious eradication that humanity is currently responsible for we can see our part in it. What follows is always a question and the question is always the same and it is:

'Is this how it is to be human?'"

– Peter Owen Jones

Turning point

"This year and next will be seen as the turning point at which the futility of governments in dealing with climate change was finally exposed."
— James Hansen, director of climate program, Earth Institute, Columbia University; NASA scientist; 2023

Strength of climate occurrences

"The climate year 2023 is nothing but shocking, in terms of the strength of climate occurrences, from heatwaves, droughts, floods, and fire, to rate of ice melt and temperature anomalies – particularly in the ocean."

– Johan Rockstrom, Potsdam Institute for Climate Impact Research, in World will look back at 2023 as year humanity exposed its inability to tackle climate crisis, scientists say, by Jonathan Watts; TheGuardian.com; Dec. 2023

2023

Scientists are sounding the alarm

"Scientists are sounding the alarm over record low levels of sea ice off the coast of Antarctica ahead of the start of spring in the Southern Hemisphere. Sea ice helps to prevent the rapid flow of ice from Antarctica's glaciers into the ocean, which can drive global sea level rise.

Satellite data shows 930,000 fewer square miles of sea ice extent surrounding Antarctica, compared to the September average – far below any previous winter level. Walter Meyer, a researcher at the National Snow and Ice Data Center told the BBC, 'It's so far outside of anything we've seen it's almost mind blowing.'

Here in New York, police arrested at least 149 climate protesters on Monday after they flooded Manhattan's financial district for a day of peaceful protests. Activists are demanding banks stop funding coal, oil, and gas projects. More protests are planned for today.

In Massachusetts, nine members of the environmental movement Extinction Rebellion were arrested Monday as they peacefully occupied the office of Governor Maura Healy to demand an immediate phase out of fossil fuels."
– Amy Goodman, Democracy Now; Sept. 2023

Protests against fossil fuels industries

And so the news went, and carried on with more protests here and there in the U.S., and in other countries.

In September, protests in Amsterdam involving an estimated 25,000 people requesting no new government subsidies for the fossil fuels industries lasted for days. The protestors blocked a major motorway. The police brought water trucks to spray the crowds, who only sat and danced in the spray on those very hot days. More than 3,000 people were arrested. But the fossil fuel executives, and the politicians who cater to the fossil fuel industry executives, weren't. (See: JustStopOil.org.)

The "do nothing: approach" isn't working

Not much of any actions are being taken to solve the problems.

At least people are speaking up. Not enough.

Considering what has been going on for decades, and how dire the situation is, it seems reasonable to expect massive protests to be taking place in all of the industrialized countries, especially around government offices, where laws are passed. And around corporate offices, where decisions are made to conduct globally ruinous investments in fossil fuels industries, the plastics industries, the meat and dairy industries, the feed crop industries, the oil palm industries, the timber industries, the GMO and chemical industries, and so forth and so on.

I thought by now humans would be far along into more sustainable practices – not only to reduce harm to Earth, but to reverse damage to wildlife habitat, and protect forests, grasslands, wetlands, rivers, lakes, marshes, and oceans.

Metal and industrialization

It hasn't been so long since the global metals industries took off. Much of the trajectory was driven by the combustion engine and the development of trains, trucks, cars, motorcycles, and airplanes. And military vehicles, ships, planes, and weaponry. There is also the industrial food industry with its massive freight vehicles and processing, refrigeration, display, and sales infrastructure. And the building of skyscrapers, bridges, and other structures reliant on metal. And also making household plumbing, stoves, refrigerators, washers, and dryers.

To make all that stuff, tremendous amounts of fossil fuels have been – and are – used. It all creates greenhouse gasses, especially CO_2 and methane. This is something humans have known for many years. That is, they've known greenhouse gasses are globally damaging.

Eunice Newton Foote and CO_2

In 1856, American scientist Eunice Newton Foote discovered that, when placed in the sun, bottles of CO_2 heat faster and to higher temperatures than do bottles of oxygen.

Foote warned that burning fossil fuels could alter the climate.

Too bad people didn't listen to Eunice Newton Foote, and act accordingly.

They've known, and failed humanity, wildlife, Nature, and Earth

People have known it. Scientists have known it. Heads of the fossil fuels industries have known it. Government leaders have known it. People working for various branches of governments all over the world have known it. It's been taught and talked about in schools all over the world, for decades. This is: The Earth's atmosphere is warming, and it is causing tremendous problems for plant, animal, and human life. And we still keep doing what causes it. That is: Burning MASSIVE amounts of fossil fuels, every second of every day, producing tremendous amounts of toxic plastics, and breeding billions of farmed animals, and growing millions upon millions of acres of crops to feed those animals. And killing tremendous numbers of wild animals. It's all a problem.

Government and corporate leaders meet, and they make agreements to get off fossil fuels, and to do things to reverse climate change. Then, nothing much is done. Whatever advances are made are wiped out by other steps backwards in the global industries and governments that continue to burn massive amounts of fossil fuels, and engage in business practices that destroy Nature – while making money for stockholders.

As I write this paragraph in 2023, the International Energy Agency is predicting oil use will increase from the current 94 million barrels per day to 102 million barrels per day in 2030. Natural gas use is also increasing. Only coal has had some decreases in use – depending on what part of the world being considered. Carbon dioxide emissions also continue to increase, and that is the main greenhouse gas.

Methane

The even stronger greenhouse gas, methane is also increasing. This is from the use of natural gas, but also from farmed animals, and from melting permafrost resulting in the formerly frozen land releasing methane. And from the heating oceans. And from the production of toxic plastics. And from deforestation.

Here we are in the year 2023 on a heating planet surrounded by increasing amounts of greenhouse gasses. As Nature dies.

The hottest year, until next year

It had been repeated in news stories all during the summer of 2023 that it had been the hottest year humans have ever experienced. At the same time, it is likely to be the coldest year humans will experience from here on out, as the temperatures will continue to rise. This is a problem unlike anything humans have ever experienced – in any century. In 2024, the story will continue.

Commercialized ignorance

With all this news, it seems relatively few people know the details about climate change, global warming, the loss of wildlife habitat, the clearcutting of the forests, the way all of it is changing the landscape, the plants, the animals, and humans. With some people, they are living in areas overwhelmed with life problems, while others can learn, but they are busy watching reality shows, sports shows, crime shows, and celebrity talk shows. They are busy shopping to replicate commercial imagery in their lives – as they drive themselves deeper into debt, which means they have to continue working at the jobs they would rather not be subjected to. The lives they live are the lives they don't want, but it is what they have worked for. They don't have time for much of anything, other than what they are already doing, caught in a turnstile of work, pay bills, spend money to replicate commercial imagery, sleep, don't get enough exercise, eat deadening, processed foods, and repeat. They have become caught in the banking system of debt, and work to pay the debt, but shop, creating more debt – while eating corporate foods. As the billionaire class gets wealthier. What are they going to do about it? With many – or most – not much. They will continue living that way. Just as they will continue living not knowing about the details of how the highly unsustainable, pollution-rich life they live is ruinous to the Nature they rely on for life. At least, they will do it… until it falls apart.

The collapse

It is likely that people this century will experience the collapse of society. Money will become worthless. Items once valued will not be. There will be no phone, TV, radio, Internet, or news services. There will be no air conditioning, heat, or mechanical venting systems. There will be no food systems or tap water. Transportation systems will not operate. Fuel and electricity will be unavailable. The government systems failed. The school, legal, police, military, and prison systems abandoned. Government buildings and businesses will be raided and burn. There will be no fire departments. Pharmacies, and medical services won't be available. The weather will be contentious, and an ongoing mystery. The remaining days will consist of uncertainty, rumors, panic, chaos, crime, struggle, exposure, thirst, hunger, and regional isolation.

The previous paragraph might describe your future, and final days.

"Civilization is, by its very nature, a long-running Ponzi scheme. It lives by robbing Nature and borrowing from the future, exploiting its hinterland until there is nothing left to exploit, after which it implodes. While it still lives, it generates a temporary and fictitious surplus that it uses to enrich and empower the few and to dispossess and dominate the many. Industrial civilization is the apotheosis and quintessence of this fatal course. A fortunate minority gains luxuries and freedoms galore, but only by slaughtering, poisoning, and exhausting creation. So we bequeath you a ruined planet that dooms you to a hardscrapple existence, or perhaps none at all."

– William Ophuls, author of *Plato's Revenge: Politics in the Age of Ecology*

Will it happen? I don't know. Could it happen? It seems to be increasingly likely. Shades of it exist. Ask the millions who lost all they own in 2023 as floods, fires, mudslides, other disasters, and wars and political upheaval took away their homes, and ended lives.

"Scientists are warming that we are now in 'uncharted territory' as a result of human-driven climate change in a new 'state of the climate' report signed by 15,000 researchers from 163 countries.

Researchers emphasized the current suffering caused by record-breaking climate extremes, and raised alarms about the possibility of widespread societal and ecological collapse in the future, while also decrying recent increases in subsidies to the fossil fuel industry, which is the primary driver of climate change.

The 2023 report, published in the journal *BioScience*, is the latest update in an annual series called World Scientists Warning of a Climate Emergency. Since 2019, scientists have been tracking escalating threats that warming global temperatures present to humans and ecosystems around the world.

The new report, led by Oregon State University Ecologist William Ripple, warns that 2023 was a particularly devastating year of extreme wildfires, floods, heat waves, and other natural disasters that are amplified by climate change."

– Becky Ferreira, *15,000 Scientists Warn Society Could Collapse This Century In Dire Climate Report*; Vice.com; Oct. 2023

"As scientists, we are increasingly being asked to tell the public the truth about the crises we face in simple and direct terms. The truth is that we are shocked by the ferocity of the extreme weather events in 2023. We are afraid of the uncharted territory that we have now entered."

– William Ripple, ecologist, Department of Forest Ecosystems and Society, Oregon State University

Record-setting hurricane leaves Acapulco in shambles

I'm writing this paragraph in October 2023, as the news is filled with stories of the damage done last night to Acapulco, Mexico by Hurricane

Otis. The storm had intensified to a Category 5 in record speed, and was the first of its strength to ever hit Mexico since weather records started being kept in the 1800s. It left the tourist-dependent city of one-million people and its homes, resorts, and commercial districts in shambles.

Hurricanes on the Pacific coast of North America have not been of concern. Until this year. The Acapulco hurricane hit months after another hurricane hit northwest Mexico in the summer of 2023, which brought tremendous rains and flooding, and transformed landscapes of Southern California's desert areas.

Land destabilization impacting homes, farms, businesses & governments

Each intense hurricane, every massive storm, more common and larger wildfires, and the destabilization of land where people live in canyons, mountains, on or near wetlands, along or on ancient lakebeds, and near lakes and rivers, and along coastlines with rising seas all increases the cost of living. All of it is what people should get used to.

"The scale of destruction from Hurricane Ian threatens to destabilize Florida's insurance and real estate markets, as devastated residents file a record number of claims for damaged or destroyed homes.

Privately insured losses from Ian are expected to reach $67 billion, not including flood insurance, according to an estimate by RMS, a catastrophe model firm.

Climate change is making hurricanes and other disasters more destructive and pushing up the cost of home insurance until it's out of reach for many people. More violent storms, flooding, and wildfires in states like Louisiana and California are causing insurers to pull back on those markets."

– Christopher Flavelle, *Why Ian May Push Florida Real Estate Out of Reach for All but the Super Rich*; The New York Times; Oct. 2022

At what point does it get too expensive and complicated to build in certain areas where hundreds of thousands – or millions – of people live? What will people do, where will they go? What will they eat?

What about the destabilization and ruin of more and more farmland that is being flooded and polluted, or where weather has gotten too hot and dry – or too wet or undependable – to grow certain crops?

I am writing this paragraph in November 2023, a year of such intense weather that there have been all of the events happening on scales never experienced. These include fires in Canada, heat scorching cropland in Europe, record flooding in major cities (including New York City), the incineration of a town in Hawaii killing many people, a damn breaking in Libya killing scores of people, storms causing landslides in India that demolished homes, thousands of acres of farmland in California polluted with toxic flood water, and scorching heat in Australia and South America. The events are too numerous to mention.

Climate emergency

In October 2023, as mentioned above, the 2019 *World Scientists Warning of a Climate Emergency* report by the Alliance of World Scientists was updated. Here are some excerpts from the document published in *BioScience* journal. Notice the number of scientists from around the world who have signed onto the report. You can read the entire document on the *BioScience* site:

"We declare, with more than 11,000 scientist signatories from around the world, clearly and unequivocally that planet Earth is facing a climate emergency.

Exactly 40 years ago, scientists from 50 nations met at the First World Climate Conference (Geneva 1979) and agreed that alarming trends for climate change made it urgently necessary to act. Since then, similar alarms have been made through the 1992 Rio Summit, the 1997 Kyoto Protocol, and the 2015 Paris Agreement, as well as scores of other global assemblies and scientists' explicit warnings of insufficient progress. Yet greenhouse gas emissions are still rapidly rising, with increasing damaging effects on the Earth's climate. An immense increase of scale in endeavors to conserve our biosphere is needed to avoid untold suffering due to the climate crisis.

Most public discussions on climate change are based on global surface temperature only, an inadequate measure to capture the breadth of human activities and the real dangers stemming from a warming planet. Policymakers and the public now urgently need access to a set of indicators that convey the effects of human activities on greenhouse emissions and the consequent impacts on climate, or environment, and society.

The climate crisis is closely linked to excessive consumption of the wealthy lifestyle. The most affluence countries are mainly responsible for the historical global greenhouse emissions. and generally have the greatest per capita emissions.

Profoundly troubling signs from human activities include sustained increases in both human and ruminant livestock populations, per capita meat production, world gross domestic product, global tree cover loss, fossil fuel consumption, the number of air passengers carries, carbon dioxide (CO_2) emissions, and per capital CO_2 emissions since 2000.

Especially disturbing are concurrent trends in the vital signs of climatic impacts. Three abundant atmospheric greenhouse gasses (CO_2, methane, and nitrous oxide) continue to increase, as does global surface temperature. Globally, ice has been rapidly disappearing, evidenced by declining trends in minimum summer Arctic sea ice, Greenland, and Antarctic ice sheets, and glacier thickness worldwide. Ocean heat content, ocean acidity, sea level, area burned in the United States, and extreme weather and associated damage costs have all been trending upward. Climate change is predicted to greatly affect marine, freshwater, and terrestrial life, from plankton and corals to fishes and forests. These issues highlight the urgent need for action.

Global daily mean temperatures never exceeded 1.5 degree Celsius above pre-industrial levels prior to 2000, and have only occasionally exceeded that number since then. However, 2023 has already seen 38 days with global average temperatures above 1.5C by September 12th – more than any other year – and the total may continue to rise.

The effects of global warming are progressively more severe, and possibilities such as a worldwide societal breakdown are feasible, and dangerously underexplored.

By the end of this century, an estimated 3 to 6 billion individuals – approximately one-third to one-half of the global population – might find themselves confined beyond the livable region, encountering severe heat, limited food availability, and elevated mortality rates because of the effects of climate change.

We warn of potential collapse of natural and socioeconomic systems in such a world where we will face unbearable heat, frequent extreme weather events, food and fresh water shortages, rising seas, more emerging diseases, and increased unrest and geopolitical conflict.

To address the overexploitation of our planet, we challenge the prevailing notion of endless growth and overconsumption by rich countries and individuals as unsustainable and unjust. Instead, we advocate for reducing resource overconsumption; reducing, reusing, and recycling waste in a more circular economy; and prioritizing human flourishing in sustainability.

As we will soon bear witness to failing to meet the Paris Agreement's aspirational 1.5°C goal, the significance of immediately curbing fossil fuel use and preventing every further 0.1°C increase in future global heating cannot be overstated.

Rather than focusing only on carbon reduction and climate change, addressing the underlying issues of ecological overshoot will give us our best shot at surviving these challenges in the long run. This is our

moment to make a profound difference for all life on Earth, and we must embrace it with unwavering courage and determination to create a legacy of change that will stand the test of time."
  – *World Scientists Warning of a Climate Emergency*, a 2019 report by the Alliance of World Scientists, signed by 15,000 scientists; *BioScience*; updated Oct. 2023

Financial implications

As in previous years, the weather continues to break records around the planet this year, causing billions upon billions of dollars in damage to many countries, including to structures, businesses, roads, and farmland.

How long can these climate change disasters continue before it impacts too many millions of people, damages too many structures, ruins too much cropland, and costs too much money for countries to keep up? It seems increasingly possible that we soon will learn.

While there are many negatives, and frighteningly awful things, I can think of some benefits to a societal collapse: no more reality stars, or celebrity gossip.

The plunge

It is likely the vast majority of people don't pay much attention to the environmental news, and even fewer to what is going on in the oceans, which they unknowingly – in many ways – rely on for things like oxygen, food, and water. It is likely they know more about celebrities than they do about things like desertification, monocropping, mountaintop removal, ocean acidification, ocean dead zones, rainforest clearcutting, the plunge in wildlife populations, pharmaceutical pollution, the problems caused by tar sands mining, the dangers of microplastics, and the complications of a warming sea.

Snow crabs

In 2023, marine biologists from the U.S. National Oceanic and Atmospheric Administration's Alaska Fishery Science Center reported that from 2018 to 2019, because of water temperature rise, an estimated 10 billion snow crabs starved to death in the Bearing Sea. The warmer water meant they needed more food to survive. But the food was not available. So, they starved, and couldn't sustain breeding.

"Billions of snow crabs have vanished from the waters near Alaska in recent years, and scientists believe the warmer ocean temperatures likely starved them to death.

The finding comes just days after the Alaska Department of Fish and Game suspended the snow crab harvest season for the second year in a row, citing an excessive quantity of crabs missing from the Bearing Seas' generally chilly, hazardous waters.

Scientists first noted a significant decrease in the quantity of snow crabs during a survey in 2021, which 'found the fewest snow crabs on the eastern Bearing shelf since the survey began in 1975,' according to the report.

Warmer ocean water is believed to have disrupted the crabs' metabolism and raised their calorie requirements.

Because of the heat disrupted much of the Bering Sea's food web, snow crabs had a difficult time hunting for food, and couldn't keep up with the caloric demand."

– Anna Louise, *10 Billion Crabs In Alaska Slowly Starved To Death Due To Extreme Marine Heat Waves*; NatureWorldNews.com; Oct. 2023

Snow crabs are one of pretty much almost every type of marine creature being impacted by climate change, ocean acidification, plastic pollution, pharmaceutical pollution, landscape and farm chemicals, algal overgrowth or undergrowth, nutrition plunge, and warming oceans.

Fish suffocation

Warmer water holds less oxygen than colder water. As more areas warm to a state of not being able to hold enough oxygen for the life there to survive, it is a condition of the water being hypoxic. This leads to hypoxia in the fish, which is a state of lacking sufficient oxygen in the tissues. Fish can detect if water has low oxygen. But if they can't get to better water, they die. More fish are dying that way, and more types of fish are being found in areas they didn't previously live – as they seek water with adequate oxygen, which usually means cooler and/or deeper water. But, because fish are also only able to live in certain depths, they can also not survive such a change of pressure. Their normal type of food – or any food – might also not be available at a different depth. They also become susceptible to a different variety of predators. It is complicated. With fish already depleted because of overfishing and destruction of habitat, all of this is leading to lower populations of fish in the oceans.

Dead zones

With warmer and more acidic oceans, and with increasing dead zones in the oceans being caused by farming, sporting field, and landscape chemicals, and farm animal, city, and industrial pollution, and massive amounts of plastic and pharmaceutical pollution, and also algal overgrowth – which leads to bacterial overgrowth as the bacteria feed off the dying algae and produce the greenhouse gasses methane and nitrous oxide – the oceans are facing numerous layers of problems that are impacting life reliant on them. Even the matter of marine life needing more calories when they are warmer causes problems, as that can lead to a decrease in their food sources. As with the sow crabs, it is a situation leading to mass starvation.

Starving whales

Whales are another ocean species with populations decreasing so quickly it is astounding, and troubling.

Malnourished whales have been washing up on the ocean shores.

Whales play major roles in the circle of life in the oceans, including by helping to spread nutrients that help other sea creatures and also marine plants to exist. The fewer marine plants there are, the less oxygen the oceans produce. As mentioned elsewhere, the oceans produce more oxygen then the land plants. We are reliant on whales.

"The lack of robust ice means less algae is growing beneath ice sheets and falling to the sea floor. This has left the whales' prey – small shrimp-like crustaceans called amphipods – in smaller numbers because of a lack of nutrients and the habitat they need. Warmer Arctic waters and faster currents have also brought the arrival of other creatures to compete for nutrients at the bottom of the food chain, further reducing the number of calorie-dense amphipods that whales need.

Between 2016 and 2023, the population of eastern North Pacific gray whales declined by nearly half, from a high of 27,000 to about 14,500 today, according to the most recent data from the National Oceanic and Atmospheric Administration."

– Alex Baumhardt, *Whales: Unusual deaths of hundreds of West Coast gray whales linked to lack of Arctic ice*; Oregon Capital Chronicle Pilot; Oct. 2023

Krill

Whales are only one of many ocean animals finding fewer sources of adequate food. Besides whales, other animals that eat krill include ice fish, penguins, seabirds, seals, and squid. Krill populations have been plunging, as have wild animals dependent on krill as food.

Humpback whale populations and the decreasing krill presence

"Research led by scientists at the University of California, Santa Cruz, shows reduced krill supplies lead to fewer pregnancies in humpback whales – a finding that could have major implications for industrial krill fishing.

The study, published January 15 in *Global Change Biology,* is based on eight years of data on humpback whale pregnancies (2013-2020) in waters along the Western Antarctic Peninsula, where krill fishing is concentrated.

Krill availability in the year before a humpback pregnancy is crucial because females need to increase their energy stores to support the upcoming pregnancy. In 2017, after a year in which krill were abundant, 86% of the humpback females sampled were pregnant. But in 2020, following a year in which krill were less plentiful, only 29% of humpback females were pregnant."

– Tim Stephens, *Study reveals influence of krill availability on humpback whale pregnancies*; UC Santa Cruz NewsCenter; News.UCSC.edu, Jan. 2023

"Krill supplies vary depending on the amount of sea ice because juvenile krill feed on algae growing on sea ice and also rely on the ice for shelter. In years with less sea ice in the winter, fewer juvenile krill survive to the following year. The impacts of climate change and likely the krill fishery are contributing to a decrease in humpback whale reproductive rates in years with less krill available for whales."

– Ari Friedlander, professor of ocean sciences, UC Santa Cruz, 2023

"The krill surplus hypothesis of unlimited resources available for Antarctic predators due to commercial whaling in the 20th century has remained largely untested since the 1970s. Rapid warming of the Western Antarctic Peninsula over the past 50 years has resulted in decreased seasonal ice cover and a reduction of krill. The latter is being exacerbated by a commercial krill fishery in the region. Despite this, humpback whale populations have increased but may be at a threshold for growth based on these human-induced changes.

Continued warming and increases fishing along the Western Antarctic Peninsula, which continue to reduce krill stocks, will likely impact this humpback whale population and other krill predators in the region.

Humpback whales are a sentinel species of ecosystem health, and changes in pregnancy rates can provide quantifiable signals of the impact of environmental change at the population level.

Our findings must be considered paramount in developing new and more restrictive conservation and management plans for the Antarctic marine ecosystem and minimizing the negative impacts of human activities in the region."

– *A surplus no more? Variation in krill availability impacts reproductive rates of Antartica baleen whales*; Global Change Biology, OnlineLibrary.wiley.com; Jan. 2023

"Krill are not an inexhaustible resource, and there is a growing overlap between industrial krill fishing and whales feeding at the same time. Humpback whales feed in the Antarctic for a handful of months a year to fuel their annual energetic needs for migration that spans thousands of kilometers. We need to tread carefully and protect this unique part of the world, which will benefit whales across their entire range."

– Chris Johnson, World Wide Fund for Nature's Protecting Whales & Dolphins Initiative

Krill populations have been declining along the Western Antarctic Peninsula since 1976. Because of warming oceans, changes to ocean currents caused by climate change, and because of loss of ice, and water acidification, krill are expected to continue declining. Perhaps at much faster rates than in previous decades. This will devastate many types of marine life, including whales – and decrease nutrient spread.

Antarctic seals and krill

"Losing fur seals from the South Shetland Islands means losing crucial genetic diversity the species may need to adapt to rapid climate change.

Research biologist Douglas Krause has studied fur seals in Antarctica for more than 20 years. As leader of pinniped studies in the Antarctic Ecosystem Research Division at NOAA Fisheries Southwest Fisheries Science Center, he and many colleagues thought they knew how climate change would affect the photogenic fur seals that are a key species in the Antarctic ecosystem.

They figured that the fur seals would benefit from warmer temperatures and habitat that would emerge as sea ice melts along coastlines. The species should flourish, they thought, a winner in the climate change lottery that affects different species in different ways.

The reality turned out to be just the opposite.

Since 2007, the number of fur seals at Cape Shirreff has decreased by 86 percent. The few pups born each year and their mothers face an uphill battle to keep the pups alive long enough to contribute to the population themselves.

The main food for the fur seals during the summer breeding season is Antarctic krill. Historically, krill were abundant around Cape Shirreff. But over the next several decades, climate change – warming ocean temperatures and declining sea ice – may reduce the amount of krill around the South Shetland Islands. Less krill puts fur seals in greater competition with other seals, baleen whales, and the fishing industry.

Without abundant krill readily available close to their breeding beaches, fur seal mothers are already having to travel farther to find food."

– An Isolated Population of Antarctic Fur Seals Could Save the Species, but They're Disappearing; Fisheries.NOAA.gov; May 2022

Antarctic seals had nearly been driven into extinction by hunters killing them so the fur can then be sold into the fashion industry. As the hunting was reduced, the seal populations began to recover. But the warming temperatures reducing the amount of krill, humans taking krill from the water, longer hunts by mother seals to find and eat krill for both their energy and to produce milk, and the babies being left in the open for longer periods of time, which makes them susceptible to be eaten by leopard seals, has all plunged the populations of the seals.

"South Georgia is in the Scotia Sea, which holds over half of the estimated global Antarctic krill biomass, is also the area with the highest regulated commercial krill extraction, and where greenhouse warming is driving increases in ocean temperatures, changes in sea-ice seasonality and extent, increases in UV radiation, and ocean acidification, affecting the marine biota. Here, the krill distribution is likely to be contracting Poleward, while increasing migration towards the seabed, both of which decrease its availability to millions of highly dependent predators such as Antarctic fur seals.

The effects of krill extraction on fur seal populations should be re-evaluated in the light of increasing environmental pressures on both the krill and the fur seals. Besides climate warming, these pressures include the effects of regional humpback whale recovery, and its direct competition for food with fur seals and other abundant krill-dependent predators such as macaroni penguins."

– Ninety years of change, from commercial extinction to recovery, range expansion and decline for Antarctic fur seals at South Georgia; OnlineLibrary.Wiley.com; July 2023

Humans unnecessarily gather krill from the oceans, including to make "krill oil supplements" – as a source of omega-3s and antioxidants.

A salty paste is made from krill in the Philippines.

Krill are also used in some Russian and Japanese dishes.

Krill meal is also used as a protein source for dogs, cattle, farmed fish, and koi fish, and "ornamental fish" kept in aquariums.

The term "ornamental fish is so warped. As if tropical and subtropical fish are meant to be part of home, hotel lobby, restaurant, and office décor, displayed in fish tanks as fashion statements. Animals don't exist for our entertainment, or to be our fashion statements. They have their own lives to live, with their families. They need healthy, natural terrain, where they can thrive and play their role in the circle of life.

Humans have no nutritional need for krill. We easily get the protein, omega-3s, antioxidants, and other nutrients from a variety of plants.

Taking too much fish and krill oil supplements sold as human-health boosters can lead to problems with blood clotting, and to a stroke. This is what happened to a friend in his 30s. He was on a health kick, and thought taking more of the fish and krill oil was better. Then he spent months in physical therapy recovering from a stroke.

Cows and dogs would never naturally eat krill. We don't need to be depopulating the oceans of krill to feed farmed or domesticated animals.

One issue with krill for human and pet foods is that krill are increasingly being found with higher and higher levels of mercury, the result of pollution. If krill are absorbing more and more mercury, the cows fed it then are exposed to the mercury, and people who eat cows then absorb some of it.

As supplement marketers have been so successful in getting humans to take krill "supplements," and with an increase in the number of fish being farmed in large enclosures, and with more krill also being added to pet food and livestock feed, the demand for krill has tremendously increased. Ships have been built or altered – including with suction harvesting tools – to gather more and more krill from the oceans, leaving fewer sources of food for wildlife. This is in addition to ocean warming that is also causing krill populations to plunge.

Krill are better off left in the ocean, where sea life depend on them.

"Krill are small crustaceans found in all oceans over the world. While krill are small individually, with an average adult length measuring six centimeters, there are 700 trillion in the Southern Ocean alone. The biggest reason for this tiny creature's steadily declining population is that the Arctic Ocean is getting warmer year after year. As water temperatures rise, glacial ice begins to melt. The krill need that ice to feed off the algae that grow under it.

Antarctic krill are at the center of the food web in the Southern Ocean, feeding multiple populations of penguins, whales, seals, and many other forms of marine life."

– Jake Cavanaugh, *Declining krill population means unhealthy oceans*; Loquitur Cabrini University Student Media; TheLoquitur.com; May 2023

Schools of krill are so massive they can be seen on satellite imagery. Images from previous years are used to monitor the global decrease in krill populations. This is partially why the ocean color has been changing. (See: *Climate Change is Shifting the Color of Earth's Oceans*, *Smithsonian Magazine*, July 24, 2023)

Please, so not support the krill oil supplement industry.

Lobsters vanishing

Just as other shellfish, including krill, lobster were once so common in the ocean they were considered a low-quality food. Lobster were so easy to gather from the ocean the meat of them was fed to prisoners, and ground up lobsters were used as fertilizer on farms and in home gardens. Lobster was such an inexpensive meat the poor were more likely than the rich to eat lobster.

As railroads were built and people began traveling by rail, lobsters from the coast were taken and kept in large tanks to then feed to the passengers. Unlike meat, living lobsters could be kept without refrigeration, and only killed when meals were made. The inland people began to favor lobster. Canned lobster started being sold. As the restaurant industry became more common, so did serving lobster as if it were some sort of delicacy. As that happened, the lobster industry expanded, with more and more traps being set, and more people depending on that income from selling lobsters as human food.

So many lobsters have been taken from the oceans the populations of them have plunged. Pollution, ocean acidification, and warmer oceans also have negatively impacted lobster populations.

"Officials with the regulatory Atlantic States Marine Fisheries Commission said surveys have detected a 39% decline in young lobsters in the Gulf of Maine and Georges Bank areas for 2020-22, compared to 2016-18. The areas are among the most important lobster fishing grounds in the world.

Scientists have raised concerns that warming waters could pose a threat to the future of the lobster industry, which produces some of the post popular seafood in the country."
– Patrick Whittle, *Young lobsters show decline off New England, and fishermen will see new rules as a result*; APNews.com; Oct. 23

Whales and carbon sequestration

"Researchers believe rebuilding populations of great whales could significantly increase the vast amount of atmospheric carbon absorbed by tiny marine algae called phytoplankton, which rely on nutrients from the leviathans' [whales'] fecal plumes.

The largest animal that has ever lived, blue whales can grow up to 100 feet long and weigh over 180 tons. An entire African elephant could fit inside the mouth of these giants. Like all living things, whales accumulate carbon in their bodies as they grow, and because they typically sink when they die, they take all of that carbon with them to their watery graves. Upon its demise, the average whale can carry the equivalent of 1,500 trees worth of carbon to the bottom of the ocean."

– Mitch Anderson, *The Planet-Saving Potential of Whale Poop*; *ReasonsToBeCheerful*; Feb. 2020

Right whales in Gulf of Maine

The right whales off the coast of New England and Canada are showing up in places uncommon to them. They are looking for food. Right are commonly found in the Gulf of Maine. The whales nearly went extinct 100 years ago, with surveys concluding there were only around 100 right whales alive. In 1935, right whale hunting was banned. In the 1970s, they were listed as endangered species. By 2010, acoustic listening stations and airplay surveys counted the population as more than 400 whales. In recent years, the right whale population has been plunging, with the population down to a little more than 300.

Among the ways right whales are dying is by getting entangled in fishing nets and ropes (from lobster and trawl fishing), colliding with ships, pollution, and the warming and acidifying oceans causing a decline in their main food source, the crustacean copepod called calanus finmarchicus. The Gulf of Maine has been identified as one of the fastest-warming ocean areas of Earth. The warming is also causing a shift in ocean circulation. Warmer water from the south is going more north, with colder waters retreating northward. This is all causing a variety of problems for all sorts of marine animals.

Whales important in the circle of life

How do you bring back whale populations when their food source is vanishing because of fishing, and the warming of seas caused by the use of burning massive amounts of fossil fuels every second of every day all over the world? It seems like a gargantuan task to change it, and to prevent the whales from going extinct.

The growth of and activities of whales help to remove massive amounts of greenhouse gasses from the atmosphere. One reason for this is because they support ocean plant life, including the growth of phytoplankton that absorb a significant amount of carbon, and produce vast quantities of oxygen. Another reason is that whales absorb carbon. When they die, the carbon is sequestered in the sea.

See the circle? Whales eat krill. Krill eat phytoplankton. Phytoplankton absorbs carbon and produce oxygen. Whales sequester carbon, and their eventual dying bodies bring it to the bottom of the ocean. This means there are many reasons why having fewer whales in the oceans – and whales going extinct – would cause enormous problems, for ocean life, and for land life, including humans.

There are currently many, many, many global problems leading to the plunge in whale populations. Among them is that Iceland, Norway, and Japan continue to hunt and kill whales. Many whales are also killed by ocean liner strikes. Some also die from plastic pollution, and, as mentioned, others die by becoming entangled in fishing nets and ropes. The largest problem is the warming oceans and loss of food for whales.

Ocean oxygen reduction and sea ice loss

"As the atmosphere warms, oceans around the world are becoming ever more deprived of oxygen, forcing many species to migrate from their usual homes. Researchers expect many places to experience a decline in species diversity, ending up with just those few species that can cope with the harsher conditions.

Lack of ecosystem diversity means lack of resilience. 'Deoxygenation is a big problem,' says University of British Columbia fisheries researcher Daniel Pauly.

Our future ocean – warmer and oxygen-deprived – will not only hold fewer kinds of fish, but also smaller, stunted fish and, to add insult to injury, more greenhouse-gas producing bacteria, scientists say. The tropics will empty as fish move to more oxygenated waters, says Pauly, and those specialist fish already living at the poles will face extinction.

Researchers complain that the oxygen problem doesn't get the attention it deserves, with ocean acidification and warming grabbing the bulk of both news headlines and academic research. Just this April, for example, headlines screamed that global surface waters were hotter than they have ever been – a shockingly balmy average of 21C. That's obviously not good for marine life. But when researchers take the time to compare the three effects – warming, acidification, and deoxygenation – the impacts of low oxygen are the worst."

– Nocola Jones, *It's Getting Harder for Fish in the Sea to Breathe: As ocean oxygen levels dip, marine ecosystems face an uncertain future*; The Tyee: Independent journalism that swims against the current; TheTyee.ca; July 2023

"Accelerated ice melt in west Antarctica is inevitable for the rest of the century, no matter how much carbon emissions are cut, research indicates. The implications for sea level rise are 'dire.' Scientists say, and mean some coastal cities may have to be abandoned.

The ice sheet of west Antarctica would put up the oceans by 5 meters if lost completely. Previous studies have suggested it is doomed to collapse over the course of centuries, but the new study shows that even drastic emissions cuts in the coming decades will not slow the melting.

Many millions of people live in coastal cities that are vulnerable to sea level rise, from New York to Mumbai to Shanghai, and more than a third of the global population lives within 62 miles (100km) of the coast."

– Damian Carrington, *Rapid ice melt in west Antarctica now inevitable, research shows: Sea level will be driven up no matter how much carbon emissions are cut, putting coastal cities in danger*; TheGuardian.com; Oct. 2023

"Our study is not great news – we may have lost control of west Antarctica ice shelf melting over the 21st century. It is one impact of climate change we are probably just about going to have to adapt to, and very likely this means some coastal communities will either have to build [defenses] or be abandoned."

– Dr. Katilin Naughton, British Antarctic Survey; 2023

Ocean fever

That is some of what has been going on with marine life and the ocean water because of the rising temperatures. The ocean has a fever.

In addition to that, people might want to know about how much they rely on coral, sharks, whales, and sea vegetation for their existence, but those forms of life are decreasing, the roles they play in the circle of life are being impacted. This marine life decline is already impacting you.

Nature decline normality

On the land, there is a tremendous decrease in plant and animal life. It is caused by the combination of climate change, urban sprawl, pollution, monocropping, livestock ranching, and killing wildlife.

Maybe some people have become desensitized to the news of environmental and wildlife issues. Or, maybe ignoring it is their way of coping and carrying on with less stress by not paying attention. Listening to music and watching fictional characters on their electronic screens can be less stressful than paying attention to environmental reality.

"Never before in its history has humanity built a civilization that had at its disposal so many different technologies to monitor, measure, and predict its own collapse – yet has been so incapable of doing anything about it.

Humans have illegally invaded, appropriated, occupied, and colonized Earth's richest habitats in order to create 'civilizations.' The resulting plague of overpopulation has brought about the climate catastrophe and collapse of all life-sustaining systems of the planet.

When a species becomes so efficient in the extinction of all other species, it is only a matter of time before it is consumed by an inferno of mutual extermination among its own members.

The only way to secure the survival of humanity, and of life on this planet, is to have significantly less humanity, and significantly more of the natural planet."

– George Tsakraklides, biologist and author of *In The Grip of Necrocapitalism: The Making and Breaking of A Psychonomy*

Human population

In 1700, the world's human population was estimated to be 610,000,000. In 1804, it was estimated to be 1,000,000,000. By 1900 that increased to 1,600,000,000. By 2000 it was about 6,148,898,975. As I'm typing this paragraph in October, 2023 the population is about 8,068,171,268. It's not slowing down. Yet. It will be, because awful environmental things are happening, and more will be – ones impacting millions more people every year by causing water and food shortages, drought, soil degradation, military conflict, and other issues.

Plants dying in heat and drought

"From wilting saguaros (cactus) in Arizona and hot-tub-like temperatures off the coast of Florida to increased heat-related hospitalizations in Europe, and agricultural losses in China, last month felt unusually hot. It was: Several teams have now confirmed that July 2023 was the hottest month in recorded history. And there's more to come."

– Jeff Tollefson, *Earth's hottest month*; Nature.com; Aug. 2023

Hotter years

At this point, it is unavoidable that we are going to be experiencing hotter years, with broader and longer-lasting droughts. The prediction is that average temperatures will rise above the pre-industrial level by more than 2.7 degrees Fahrenheit (1.5 degrees Celsius). We are on the pathway to it. The environmental consequences are happening faster than scientists had expected even a few years ago. As it continues, we will witness catastrophic environmental changes, and the extinction of more plant and animal species.

Eating animals

Humans continue one of the most destructive practices: consuming diets rich in animal protein. Most is from industrial farming of mammals, birds, and fish. I requires millions of acres of land to be monocropped with feed crops, uses massive amounts of water, and requires enormous amounts of fossil fuels, metals, and other resources.

One of the things scientists from around the world are strongly advising is that people get off of animal protein, follow a plant-based diet, and allow the cropland currently used for growing feed for farmed animals to revert back to wildland.

Scientist's warming

"We must protect and restore Earth's ecosystems. Phytoplankton, coral reefs, forests, savannas, grasslands, wetlands, peatlands, soils, mangroves, and sea grasses contribute greatly to sequestration of atmospheric CO_2. Marine and terrestrial plants, animals, and microorganisms play significant roles in carbon and nutrient cycling and storage. We need to quickly curtail habitat and biodiversity loss, protecting the remaining primary and intact forests, especially those with high carbon stores and other forests with the capacity to rapidly sequester carbon (proforestation), while increasing reforestation and afforestation where appropriate at enormous scales. Although available land may be limiting in places, up to a third of emissions reductions needed by 2030 for the Paris agreement could be obtained with these natural climate solutions.

Eating mostly plant-based foods while reducing the global consumption of animal products – especially ruminant livestock – can improve human health and significantly lower greenhouse gas emissions. Moreover, this will free up croplands for growing much-needed human plant food instead of livestock feed, while releasing some grazing land to support natural climate solutions. Cropping practices such as minimum tillage that increase soil carbon are vitally important. We need to drastically reduce the enormous amount of food waste around the world."

– *World Scientists Warning of a Climate Emergency*, a 2019 report by the Alliance of World Scientists, signed by 15,000 scientists; *BioScience*; updated Oct. 2023

There are many, many reasons to follow a plant-based diet, including to reduce water use, including from underground aquifers that are being sucked dry and polluted by the animal farming and feed crop industries.

Water, drought, dairy, and feed crops

The dairy and farmed animal feed crop industries use VASTLY more water than lawns, sporting fields, and golf courses combined. They also export billions of gallons of water from California in the forms of dairy products (milk, cheese, butter, cream, whey, casein, yogurt, kefir, ghee), and hay, soybeans, oats, and other farmed animal feed sent as far as the Middle East.

The export of dairy and feed crops involves billions of gallons of water leaving a drought-stricken region. It might make money for corporations, but it is another branch of ecocide capitalism.

California dairy & feed crop industries: water-reliant in drought region

It's odd that the dairy industry is so prominent in California, a region of long-term drought. Each dairy cow takes about 25 gallons of water per day to hydrate – more when it's hot outside. Plus all the thousands of gallons of water it takes to grow their feed.

The state and federal governments give the California dairy and feed crop industries money to market their products in other states, and in other countries. This benefits few in the state, because most of the dairy and feed crop industries are owned and controlled by corporations traded on the stock market.

The dairy industry and feed crop industry are also reliant on cheap labor who work hard, largely do not get benefits, and make minimum wage. Their resident status can also cause problems for them.

Incarcerated innocence

After a life of having been repeatedly impregnated to keep them producing milk, dairy cows are sent to slaughter to make cheap meat – which ends up in places like fast food restaurants, convenience stores, food trucks, and school, hospital, nursing home, military, prison, and other industrial kitchens, and in grocery stores.

Baby cows (females) are kept to impregnate so they produce more milk to turn into dairy products.

Baby bulls (males) are killed and their muscles sold as veal and pet food. The stomach lining – rennet – is taken and used to curdle milk in the cheese-making process. That's why cheese can smell like vomit – because it contains the stomach lining of slaughtered baby bulls.

Water conservation and drought

Here in California, we are told to be concerned about our water use. We are to take shorter showers, shut off the faucet while we brush our teeth, wash full loads of laundry, and reduce our outside water use.

What we are not told is how much water the animal farming industry uses to keep millions of farmed animals fed and hydrated. Or how much water is used by golf courses, polo and other sports fields, or corporate campuses, or church lawns. Or how much water is used to produce what is then exported, including products containing massive amounts of water. That is, dairy products, meat, and wine, and also the feed crops shipped to other countries to feed their farmed animals.

Fracking and water usage... pollution

Then there is the fracking done by the oil and gas industry in California, which uses – and pollutes – a large amount of water. According to the Pacific Institute, in California, fracking uses about 290 million gallons of water per year. In Kern County, about 20% of water use is for fracking. Other states use even more water, as their wells are often deeper, including in Texas, where about 4.25 million gallons are used for fracking, and the cattle and feed crop industries have depleted the aquafers. Texas is another drought area, with crop failures, record high heat, and water shortages. Much of the water used for fracking is then polluted with chemicals, including from the gels used in fracking.

In California, billions of gallons of water – about ten times more than what is being used for fracking – are also used to inject steam and hot water into underground oil deposits, pushing the oil to wells. That greatly magnifies the amount of water being used by the fossil fuels industries in California, home of the long-term drought. In other states, the same thing is being done.

Water, livestock, and fossil fuels

The livestock, feed crop, and fossil fuels industry leaders have the science presented to them concluding their industries are main causes of global climate change, water pollution, ocean degradation, degraded landscapes, topsoil loss, and the extinction of species. Government leaders know, and yet they still make decisions helping the fossil fuels and livestock industries. What do the companies do about it? Invest more money into their industries, including to satisfy stockholders. It's greed. They also fund what works against environmental protections, and pressure law enforcement to arrest peaceful protestors. It is insanity. It is crimes against humanity and Nature. It is ecocide.

Scientists from around the world are warning that the ocean currents are likely to shut down within not many years, which will be globally ruinous, and lead to the extinction of many species.

The human extermination of species

At this point, human activity is essentially exterminating species. Just as bug extermination chemicals also poison humans and lead to human diseases and deaths, what humans are doing to the planet's wildlife and environment is doing the same to humans.

23

Methane, CO_2, and a heating planet

Combined with massive amounts of methane from animal farming (hundreds of millions of cattle, pigs, goats, and other animals bred every year to raise for dairy, eggs, and/or meat) and growing more and more feed crops on billions of acres of monocultured land cleared of native trees and other vegetation, and with wild animals killed to protect the crops and the livestock, and tremendous amounts of increasing methane ending up in the atmosphere from permafrost regions melting and oceans heating because of the CO_2 and methane heating the planet, the evidence seems clear: humans are leading themselves into extinction.

Hot ocean and hurricanes

During the summer of 2023, the ocean off Florida was recorded as having the highest ocean temperature ever recorded anywhere on Earth. It had already been over 15 degrees above normal. It very likely will get even hotter in coming years – which is devastating to wildlife.

Hurricanes caused by this warming will be worse than we have ever seen. Storms that will continue to break records for rain, wind, ocean surge, and destruction from the coasts and far inland for hundreds of miles as the tropical storms spur floods and tornadoes.

Heating planet

We are in a situation never experienced by humans, including heat, soil loss, pollution, decreased numbers of species, the spread of desertification, massive industrial and plastic pollution, ocean acidification, melting icescapes, and other scenarios mixing together to formulate a planet that is increasingly unlivable for humans.

As I write this, I'm near the Pacific coast in Southern California. Temperatures reached the mid-80s here. Inland, temperatures are in the 100s.

Not too far away, in Phoenix, Arizona, they have had over 30 days of temperatures in the 100s.

European friends are telling me the heat there is intolerable.

As I'm writing this paragraph in the summer of 2023, crops in regions of Italy are being scorched and are dying from the heat.

Storms

Other European regions have experienced record-setting storms that have flooded streets, homes, schools, and businesses.

Recent storms in India have washed away roads, bridges, houses, and people, destroyed crops, and degraded farmland.

Fires incinerating wildlife

All this summer while I'm writing this part of the book, there have been enormous forest fires burning in Canada. The drifting smoke has smogged cities as far as the Southern U.S., and into Europe. Nobody knows – and there is no way of knowing – how many animals are dying in the fires, or how many trees, bushes, and plants are being incinerated.

"'We don't' have a precise idea of the number of animals that died, but it's in the hundreds of thousands,' says Annie Langlois, a biologist for the Canadian Wildlife Federation.

Beavers, coyotes, skunks, wolverines, foxes, bears – the Canadian boreal forest is home to 85 species of mammals, 130 of fish, and 300 of birds, including many migratory birds.

But it has been devastated by this year's record wildfire season, with more than 18 million hectares burned – an area close to the size of Tunisia."
– Mathiew Leiser and Genevieve Normand, Canada's wildfires take devastating toll on wildlife; phys.org; Oct. 2023

Also killed in the fires were unknown numbers of raccoons, caribou, moose, lynx, bats, beetles, butterflies, bees, wasps, praying mantis, spiders, and also slugs, snails, and other mollusks in the wet areas, and also snakes and turtles, and also frogs, toads, and other amphibians, and fish in the streams and rivers, ponds, and lakes poisoned by massive amounts of smoke, ash, and mud. This all adds up to many millions of animals killed by the 2023 fires in Canada. All of those animals play roles in the circle of wildlife, including by helping simply by their daily functions to support plant life and soil organisms.

Torrential rain

Every week in 2023 there has been a new story of a severe rainstorm causing flooding in a different area of the planet, from China to India to Spain and other parts of Western Europe to Australia to Brazil to Vermont and New York City and California. The rains have been demolishing buildings and homes, rerouting rivers, refilling lakes, and damaging, or obliterating roads, bridges, and dams. The rains have been flooding or completely destroying farmland, vineyards, and orchards.

Floods in Greece

During the first week of September, an estimated three-years' worth of rain fell on Greece. Because of land damage, infrastructure destruction, and the flooding of a large swath of land that was once a lake, but had been converted to farmland, and flooding other agricultural land with polluted water, and so many people lost their homes and businesses, the country suddenly lost a fourth of its capacity to grow food. The heat and storms have damaged the olive trees, ruining the fruit.

Because of the 2023 heat and storm damage to olive orchards throughout Europe, the cost of olives and olive oil has increased.

Floods in Libya

In Derna, Libya, the September storms caused a dam to collapse, sending a wall of water rushing down a canyon and through the port city, killing more than 4,000 people, and leaving thousands missing – likely washed out to the Mediterranean Sea.

Floods, and more floods

The week I'm writing this paragraph in October 2023, there have been floods everywhere from Ireland, England, to China and Thailand.

It is predicted this winter will bring more record rainstorms – and floods – to California. Last winter many California farms were blooded with polluted water. That land won't be useable for two or more years.

Storms around the world have been accompanied by hail, which has shattered vehicle windows, damaged structures, and ruined crops.

Last winter my vegetable garden mostly turned to mud under the relentless rainstorms that brought more precipitation to Southern California than Washington State had. It was far from normal.

Lack of bees and pollinators

This past summer, there were few bees, wasps, or other insects in my vegetable garden. The tomato and cucumber plants barley produced. The zucchini, eggplant, pumpkin, and pepper plants died without fruiting. Other plants were so small they weren't worth harvesting. This is bad. Friends who garden are having similar experiences. The loss of pollinator insects is caused by local use of bug sprays, and pollution, and climate change.

"The overlooked degradation of butterflies, beetles, bees, ants, ladybugs, and countless other species have huge ripple effects across our local and global ecological functions – from loss of bird populations to a reduced ability to grow food. Why are we not more concerned about the health and vitality of these critical organisms? Can humans – or life as we know it – survive without these little creatures?"

– Nate Hagens, interviewing Nick Haddad, author of *Insects: A Silent Extinction*; Resiliance.org, Sept. 2023

Insecticides

"Populations of pollinating insects are declining dramatically. For instance, flying insects have decreased by over 75% in Germany over the last 27 years. Diminishing pollinator numbers are a serious threat to our food security with intensive use of insecticides being implicated in these losses.

The most common insecticides worldwide are neonicotinoids, which account for 24% of the global insecticide market valued at $1 billion per year. Neonicotinoids are highly efficacious insecticides, however they lack specificity, affecting both target pest species, such as aphids, and non-target beneficial insects, such as bees. They share a mechanism of action, being agnonits of nicotinic acetylcholine receptors (nAChR), the main neurotransmitter in the insect nervous system. They also display target site cross-resistance in pests, diminishing their effectiveness as insecticides, and unfortunately encouraging application of increasing concentrations. They were branded safe compared to previous insecticides, and because they have little effect on mammalian nAChRs. However, few precursive safety tests were performed/published on beneficial insects, for which neonicotinoids are now known to be potent neurotoxins with well-documented lethal and sub-lethal effects. Therefore, continued intensive use is likely to have severe consequences on insect species numbers, with knock-on as well as direct effects on the ecosystem, aquatic life, birds and mammals, including potential toxicity to humans."

– *Neonicotinoids disrupt memory, circadian behavior and sleep; Scientific Reports,* Nature.com; Jan. 2021

Local microlife decline

My garden used to have snails. I haven't seen any in years. There are few worms, and I've had to buy worms to repopulate the soil.

The property in front of my place where I've planted dozens of trees used to have worms, and birds showing up to dig them. I haven't found a worm there in years. The birds who fed on them no longer show up.

I used to see snail and worm trails on the sidewalk after a rain, or even a foggy night. I haven't seen the trails in years.

I've seen few butterflies. Globally, butterflies, other insects, and birds are declining 2% per year. Humans can't exist without them.

There used to be a pair of falcon in a nearby oak tree. Year after year the falcons would be there. Sometimes they would sit on the fence and watch me in my garden. I haven't seen them in three years. Maybe they were poisoned by eating a rat who had eaten poison.

There used to be lizards in the neighborhood. On hot days the lizards sunbathed on the rocks. One particularly large lizard lived out front for years. I haven't seen a lizard here in at least six years.

I'm not sure what is happening, but something isn't right. It is climate breakdown, global warming, and a plunge in wildlife populations. Especially of the birds, bees, butterflies, and microlife.

Global boiling and ecocide

On social media the term "global boiling" has trended all year. Meanwhile the ecocide continues.

Petroleum companies continue to drill new wells and build oil pipelines.

Gas companies keep fracking.

Coal companies keep mining.

Coal-fired electric generating power plants continue to be built and used, spreading mercury poisoning across the globe. Because of this, fish are increasingly being found with mercury in their tissues. That is spread into the marine birds, and into the animals that feed on the fish that swim up into rivers to spawn.

Plastic pollution

Plastic made of petroleum and natural gas continues being made, and "thrown away" to… here, on Earth… where there is no "away."

As plastic is essentially strangling many types of wildlife, a tremendously enormous and immensely polluting plastic "cracker plant" factory recently opened in Pennsylvania on the banks of the Ohio River. The factory, owned by Royal Dutch Shell, takes up 386 acres, is fed by ethane gas (potentially leaky and explosive) pipelines from around the state. The factory is to produce more than a million tons of plastic, and emit as much carbon dioxide per year as nearly half a million automobiles. It was sold as being an economic benefit to the region, but it is a pollution monster to Earth.

To get the plastic factory built, the state of Pennsylvania offered the company tax breaks that were estimated to be valued around $1.6 billion. It is only part of what the state has been offering petrochemical companies to set up shop in the state with an agreement they will use fossil fuels taken from land in the state.

To make the polyethylene plastic, the Royal Dutch Shell plant uses ethane, a natural gas byproduct released during fracking – the tremendously polluting fossil fuel extraction process.

Fracking and its relation to chemicals gathering in your body

Here is a little info about the filthy gas industry in Pennsylvania. You can Google this to learn more – including how it is being done near you, and what it is doing to the environment, and your health.

"Oil and gas producers in Pennsylvania used some 160 million pounds of chemicals they are not required by law to publicly identify in more than 5,000 gas wells between 2012 and 2022, according to research published on Tuesday.

The chemicals may have included per- and polyfluoroalkyl substances (PFAS), a toxic and pervasive class of chemicals, according to the report from Physicians for Social Responsibility (PSR), an activist group that last week co-published a new compilation of studies on the harms of hydraulic fracturing for oil and gas.

PFAS – dubbed 'forever chemicals' because they don't' break down in the environment – and accumulate in the blood of virtually every American – are linked to serious illnesses, including some cancers low birth weights, ulcerative colitis, reduced receptiveness to vaccines, and elevated cholesterol."

– Jon Hurdle, *Pennsylvania's Gas Industry Used 160 Million Pounds of Secret Chemicals From 2012 to 2022, a New Report Says; Inside Climate News*; InsideClimateNews.org; Oct. 2023

The toxins from fracking have increasingly been found in streams and rivers, and everywhere from wells and lakes in Pennsylvania.

If you think this is only going on in Pennsylvania, you're wrong. Fracking and the filthy chemicals used for doing it are used in a growing number of regions in North America, and in other continents. This is another reason to get off of fossil fuels.

Fracking and plastic

Fracking of the Marcellus Shale in Pennsylvania and Ohio has been poisoning land for decades. As awful as it is for the environment, wildlife, and humans, and as much water pollution as it causes, fracking has only been increasing. Polluting more land, water, and air, and releasing more greenhouse gasses into the atmosphere.

"When completed, the facility will be fed by pipelines stretching hundreds of miles across Appalachia. It will have its own rail system with 3,300 freight cars. And it will produce more than a million tons each year of something that many people argue the world needs less of: plastic.

As concern grows about plastic debris in the oceans, and recycling continues to falter in the United States, the production of new plastic is booming. The plant Royal Dutch Shell is building about 25 miles northwest of Pittsburgh will create tiny pellets that can be turned into items like phone cases, auto parts, and food packaging, all of which will be around long after they have served their purpose.

The plant is one of more than a dozen being built – or have been proposed – around the world by petrochemical companies like Exxon Mobil and Dow, including several in nearby Ohio and West Virginia, and the Gulf Coast."

– Michael Corkery, *A Giant Factory Rises to Make a Product Filling up the World: Plastic*; *New York Times*; Aug. 12, 2019

More plastic pollution

More and more plastic ends up in the environment. Plastic cups, bottles, bags, packaging, discarded toys, "disposable" diapers, medical waste, and other forms of plastic are discarded, and end up as pollution.

The plastics industry has lied and lied, and lied some more about recycling. Only a small fraction of plastic is recycled. Even when it is, it eventually ends up as trash in one way or another.

One solution is to make compostable plastics out of plants like corn, potatoes, vegetable farming waste, and from industrial hemp. (Search: Compostable plastics, and hemp plastics.)

As I walk and jog on the beach, I see so much plastic pollution it would take teams of people to walk along the sand to gather it all. Sometimes I walk for miles on the beach while picking up plastic and other trash. Much of it is too small to collect as it has broken down into microplastics that will continue to cause problems for Nature and wildlife for hundreds – or thousands – of years.

Plastic fibers

"Studies have detected plastic fibers everywhere – in the stomachs of sperm whales, in tap water, and in table salt. A researcher in Britain says plastic may help define the most recent layer of the Earth's crust because it takes so long to break down, and there is so much of it."

– Michael Corkery, *A Giant Factory Rises to Make a Product Filling up the World: Plastic*; *New York Times*; Aug. 12, 2019

We don't know the long-term implications of plastic pollution

"Plastic doesn't really go away. It just accumulates and ends up in the wrong places. And we simply don't know the long-term implications of having all this plastic everywhere in the natural environment. It is like

this giant global experiment, and we can't simply pull the plug if it goes wrong."
– Roland Geyer, professor of industrial ecology, University of California, Santa Barbara

Polyethylene plastic

To produce polyethylene plastic requires using enormous amounts of natural gas to heat the ethane gas from fracking wells in an large furnace. As it cools, rice-like pellets are formed. The process is called "cracking" as it fractures molecules to make the plastic. That is why the factories are called "cracker plants." The process creates a tremendous amount of air pollution, including greenhouse gasses.

We have a situation taking place at plastic plants in many regions of the world using fossil fuels to create a toxic substance that remains a problem on Earth for an unknown amount of time.

Beach trash

Even when I have gathered bags and bags of trash from the beach, I know that in another day, more washes up on shore, and more humans toss more trash on the sand, and on city streets. Much of the street trash ends up in storm drains, rivers, and washes, and then in the ocean.

Who raised such people who think it is okay to toss trash on the beach, or out their car windows, or anywhere?

Closer

What are humans doing to Earth and wildlife?

It's too complicated to include it all in one book. But this should help give a person an idea of how Earth is in such a sickly state that, every year, humanity gets closer to extinction.

Situation is dire

"We are currently losing species at a rate far higher than normal background extinction rates, and the situation is dire. We are rapidly approaching a loss of diversity similar to that seen during mass extinctions. Biologists predict that unless we change course and begin preserving more species, within the next few hundred years, we will become the cause of Earth's sixth mass extinction."
– The Earth's sixth mass extinction?; Understanding Evolution; Evolution.Berkeley.edu; 2023

"Next few hundred years" might be a brighter outlook than what is actually going on. Especially considering the global climate and ocean changes, and species loss in the past year.

"We're in this really unusual position, where, for the first time, we are trying to put our finger on a geologically superlative event while it's happening. Just evoking the fact that our influences could even be on the scale of a comet, or some other big events in the past – I think that should be giving us pause."
– Jacquelyn Gill, paleoecologist, University of Maine

The six extinction events

The Ordovician-Silurian mass extinction
Cause: Cooling climate and sea level plunge as glaciers formed.

Late Devonian extinction
Cause: Environmental changes.

Permian-Triassic extinction
Cause: Possibly volcanic activity.

Triassic-Jurassic extinction
Cause: Volcanic activity.

Cretaceous-Paleogene extinction
Cause: An asteroid collided with Earth. Perhaps assisted by volcanic activity.

The Sixth Great Extinction: The Holocene extinction.
Cause: Human activity.

This is where you and all the rest of us exist, in an uncertain time with the environment rapidly changing, and impacting plant and animal life, the water, land, and air, and our food in ways in which humans have never experienced.

A clear indication we have entered a sixth mass extinction

"Many estimates for today's extinction rates are based on the Red List compiled by the International Union for Conservation of Nature (IUCN). Although it's the most comprehensive database of the conservation statuses of wild species, only around 142,500 animal, plant, and fungal species have been assessed – just 6.5 percent of the roughly 2.2 million species known to science, which itself is a fraction of the species on Earth. There are, however, good records for vertebrates, such as mammals and birds, for which the IUCN has evaluated nearly all described species. In 2015, based on vertebrate data, ecologist and conservation biologist Gerardo Ceballos of the Universidad Nacional Autonoma de Mexico and his colleagues estimated that the rate of species loss over the past century is up to 200 extinctions per MSY –

'incredibly high compared to what happened in the last few million years,' Ceballos says. 'For us, it was a clear indication that we have entered a sixth mass extinction."
— Katarina Zimmer, *Are We in the Midst of a Sixth Mass Extinction?*; *The Scientist*; The-Scientist.com; July 2022

The unknown of the unknown

Many types of plants, animals, and fungi have never been identified or studied – or even known of – before they went extinct in the past few hundred years. The situation has only been accelerating as the climate changes, temperatures rise, the ocean acidity increases, the permafrost and ice melts, landscapes are altered, and pollution spreads.

There are plants, animals, and fungi everywhere from remote islands to the mountains and beneath the sea that are still unknown. There are plants and animals existing in small areas, in micro-ranges. One volcanic area. One ocean bay, shore, marsh, or lagoon, or a deep-sea or lake or otherwise a marine realm location. One mountain, one canyon, one cave, or desert or forest or meadow. Some of the animals are small, undetected by people even when people are in the presence of the creatures. In this way, humans are not aware of what is out there, what is endangered, the benefits of the life forms, and what their symbiosis is, and when or why the life vanishes from existence.

(Search: Animal Welfare Institute: List of Endangered Species.)

"Most scientists agree that overall extinction rates are much higher than at any time in the past several million years, and perhaps even higher than during previous mass extinctions."
— Katarina Zimmer, *Are We in the Midst of a Sixth Mass Extinction?*; *The Scientist*; The-Scientist.com; July 2022

Whatever it is that is going on, perhaps it is time for humanity as a whole, including governments, politicians, corporate leaders, education institutions, and others to step back and learn about the possibilities of what humanity has been doing to the planet and all life on her since the beginning of the industrial revolution, and what can be done to help restore Nature, protect plants and animals, and to get humanity to live more sustainably.

"I find it hard to exaggerate the peril. This is the new extinction, and we are half way through it. We are in terrible, terrible trouble, and the longer we wait to do something about it, the worse it is going to get."
— Sir David Attenborough

Humanity needs to take action to save Nature. And it needs to do it immediately.

The number one cause of environmental breakdown, climate change, and biodiversity loss is human's food choices, and how and where humans grow, harvest, package, transport, sell, and prepare their foods.

Biggest human-caused threats

"Unlike previous extinction events caused by natural phenomena, the sixth mass extinction is driven by human activity, primarily (though not limited to) the unsustainable use of land, water, and energy use and climate change. Currently, 40% of all land has been converted for food production. Agriculture is also responsible for 90% of global deforestation, and accounts for 70% of the planet's freshwater use, devastating the species that inhabit those places by significantly altering their habitats. It's evidence that where and how food is produced is one of the biggest human-caused threats to species extinction and our ecosystems. To make matters worse, unsustainable food production and consumption are significant contributors to greenhouse gas emissions that are causing atmospheric temperatures to rise, wreaking havoc across the globe. The climate crisis is causing everything from severe droughts to more frequent and intense storms. It also exacerbates the challenges associated with food production that stress species, while creating conditions that make their habitats inhospitable. Increased droughts and floods have made it more difficult to maintain crops and produce sufficient food in some regions. The intertwined relationships among the food systems, climate change, and biodiversity loss are placing immense pressure on our planet"
– WorldWidlife.org

"Things are going to get worse. The question is how much worse, and how quickly is it going to get worse. The speed is accelerating. Whatever we do now, it's going to get worse. And unless we act within the next ten years, I mean, we are in real trouble."
– Sir David Attenborough

The Reality

Destroying Earth at a rate comparable to an asteroid collision

"We're destroying the Earth at a rate comparable with the impact of a giant asteroid slamming into the planet, or even a shower of vast heavenly bodies."
– Paleoanthropologist Dr. Richard Leakey

Nonfiction

I wish this were a book of fiction, and all of the troubled forests, meadows, plants, animals, swamps, ponds, rivers, lakes, marshes, grasslands, oceans, and land I write about in this book were some figment of my imagination. Or, at least I could magically reverse all of the damage humanity has done to the planet by way of their extracting and burning fossil fuels and breeding, feeding, killing, and cooking hundreds of billions of farmed animals in the past two hundred years.

"The current extinction event is due to human activity, paving the planet, creating pollution, many of the things that we are doing today. The Earth might well lose half of its species in our lifetime. We want to know which ones deserve the highest priority for conservation."
– Bradley Cardinale, co-author of study on extinction conducted by the University of California, Santa Barbara

Collapse

Unfortunately, this book covers real issues and the reality of what many scientists, environmentalists, and activists see as the collapse and destruction of life on Earth, including from the depths of the oceans, and throughout the landscapes, to the tops of the mountains.

If you think what happens at the extreme depths and heights of the planet doesn't impact you, it is time to reconsider that concept. Because your life depends on all of it, and the lives and interactions of wildlife.

"We have a population of seven billion people on the planet right now, and the oceans are dying. The oceans have been so severely diminished that there's a good chance we could kill them. And if the oceans die, we die."
– Paul Watson, Sea Shepherd Conservation Society founder, to *Earth Island Journal*, Autumn 2010

Fewer

As I'm writing this in 2023, the world's human population is somewhere over eight billion (nobody knows the exact figure). As there are more people, there are fewer other animals on the land, in the trees, and in the ponds, marshes, swamps, rivers, lakes, and oceans.

Every part of the continents, islands, and oceans continue to be in worsening shape, with fewer varieties of life on and in them.

Magic oceans

To me, oceans are magic. The energy of them, and the life in and around them are like a different world from the one where I grew up in the Midwest. With the oceans, you never know when schools of fish or an animal many times larger than you can suddenly appear.

When I was 31, I still had never seen a whale. I had spent loads of time in and around the Pacific and Atlantic Oceans, and I wondered where these whales were. I almost felt cheated I had never seen one. To me, when someone mentioned they had seen a whale, it was as if they were speaking of a mystical occurrence, or a dream.

Whale breach

One sunny afternoon, I was standing on the balcony of a home overlooking the ocean in Malibu. I had my girlfriend's two-year-old son in my arms and I was looking out at the rocks jutting from the water just beyond the surf. To entertain myself, I started to say to Charlie, "Look, there's a whale." Before I said the word, "whale," and as soon as my finger pointed out to one of the rocks in the ocean, a whale breached from the water just beyond the rocks, fell back onto its side, and vanished below the water.

Astounded, my breath left me as I fell back against the glass door of the balcony as I nearly dropped Charlie. It felt as if my heart had jumped into my neck. I was somewhere between laughter, tears, and euphoria. Charlie and I stood watching the whale breach more times.

Shortly, there were dozens of people running outside to watch as the whale breached from the water a few more times. Then it was gone, likely on its way to northern waters for the summer.

Dolphin jump

Ten years prior, I was living in Florida and was driving a big old tank of a car I had bought for next to nothing. It was a bright, sunshiny afternoon when I was crossing a bridge from Palm Beach to West Palm Beach. I glanced out just in time to see a dolphin jump from the water, do a slight spin, and vanish back into the waterway. It was the first time I had seen a dolphin.

There was a loud BOOM!

My car had bumped into the car in front of me.

Luckily, the old couple was driving the same model of car I had, and the frames of the cars matched up perfectly. There were no dents on either car.

The old couple seemed entertained by it all, and luckily for me, they didn't want my insurance information or driver's license number. The kind old man got back in this car and they went on their way.

Snorkeling Florida

A few weeks later, I went down to the Keys with a group of friends for a weekend of camping, swimming, boating, and snorkeling.

I had heard it is best to stay with your group while out in the ocean snorkeling.

I got a bit carried away with the beauty of the coral reefs as I drifted away from the group.

Suddenly, many thousands of small fish were passing by me. I couldn't see anything else in the water other than the fish. As soon as they passed, another school of slightly larger fish swam all around me. When they passed, the water was clear and I was able to see down to the bottom, about 30 feet below.

A shark larger than me was swimming directly beneath me. It was the first time I had seen a shark.

I looked out of the water to see the boat was about two hundred feet away, and my friends were so involved in snorkeling they didn't seem to notice me.

I calmly swam toward the boat, keeping my eye on the shark until it was out of sight. As I swam, a barracuda swam several feet alongside me. I didn't know much about barracudas, but I knew I didn't feel comfortable swimming next to a large fish with teeth that looked as if they could tear into my skin.

I was glad to get back to my friends and the boat with nothing more than a story to tell.

"Surfing soothes me, it's always been a kind of Zen experience for me. The ocean is so magnificent, peaceful, and awesome. The rest of the world disappears for me when I'm on a wave."
– Paul Walker

Sharks

A couple of years ago I had gone for a jog in the sand during sunrise here in California. After a few miles, I was as sweaty as could be. Jumping into the water, I felt invigorated swimming in the calm surf.

An uncomfortable feeling overcame me that I was not swimming alone.

No people were around, but I still felt stupid running out of the water as if I were being chased.

I didn't know what it was that gave me the uncomfortable feeling, and I was glad nobody saw me run from the water as if I were afraid of my own shadow.

The next day there was a photograph in the newspaper of a large shark jumping from the water very close to where I had been swimming alone the previous morning. It was the first I had heard of sharks near that beach.

I don't think I will be jumping into the ocean in the morning anymore, as it is feeding time for sharks. I don't want to feed the wild animals, especially with my flesh.

I don't want to instill a fear of sharks. The chances of being bitten are extremely slim, and being seriously injured by a shark is vastly far less likely than winning the lotto. You have even less of a chance of being killed by a shark.

A little about sharks...

Things we are reliant on for air: trees and plants, soil organisms (certain types of bacteria and mycorrhizal fungi that work in combination with endophytes in the plants), pollinating animals (birds, bees, bats, butterflies, beetles, wasps), animals that spread nutrients (all animals), and water life, including river and lake creatures, coral reefs, sea mammals, shellfish, and fish, including... sharks. This is the web of life that, through their natural behavior and existence – and eventual nutrient spread through decomposition – supports all life. Almost every form of life, from the mountains to the bottoms of the seas, is in steep decline.

Sharks help fertilize the oceans and produce oxygen

Sharks play a major role in other forms of life in the oceans, including by moving large schools of fish around, which fertilizes the seas. The marine plant life is then fertilized and produces most of the oxygen you breathe. This all helps other marine life survive, resulting in food for marine birds, and for land animals that feed on marine life, which then helps spread nutrients into soil, supporting plant life.

On land, animals distribute nutrients through their natural behaviors. It is similar in the oceans. Just as the behaviors of land animals help to fertilize the soil that feeds the plants the animals rely on for food, ocean life helps fertilize the water and surface, which leads to more life that supports their life.

Just as it is important to our life for land animals survive, and thrive, it is also important that sharks and other sea life survives, and thrives.

Murdering sharks is killing Nature

Humans kill an estimated 100 million sharks per year. Many are killed because their fins are cut off to sell into the international market for shark fins, including for shark fin soup. Many other sharks are killed by humans who consider sharks to be dangerous, and they think fewer sharks is better. Those humans are dead wrong.

With that reasoning, people may as well be killing hundreds of millions of people per year.

You're greatly more likely to be harmed or killed by a human than by a shark.

Sharks help you to breathe

On the other hand, sharks help you to live. This is so no matter where you live – even if you live in the mountains.

If you breathe oxygen, thank a shark.

If you breathe pollution, blame humans.

You are intrinsically dependent on healthy oceans, and abundant populations of marine life, including sharks.

"After a visit to the beach, it's hard to believe that we live in a material world."
– Pam Shaw

The blue wall

Since I was a little boy I had wanted to live near the ocean. When I was sixteen and felt the need to leave the horrible situation I grew up in, I hitchhiked from Ohio to California. It was an amazing adventure.

Because my childhood was so troubled, I thought my life would be short. The one thing I wanted to see before I died was the ocean.

I will always remember the moment I first saw the Pacific. As the car I was in descended a hill, I became confused as there appeared to be a tremendous wall of blue topped by sky. I asked, "What is that wall?" I had never seen anything so gigantic. It took me a moment to realize I was seeing the vast, flat ocean, all the way out to the horizon. Seeing surfers in the Pacific during December when I had left freezing cold in Ohio two days before was what made me want to move to where I could enjoy the ocean all year.

Marine animal

On the hitchhiking trip I slept near some random beaches and spent part of one night watching seabirds run along the sand in the darkness. They would quickly nibble at some mysterious something they were digging out of the sand as the surf retreated. The next day I went for a long walk along the beach in Laguna and was amazed by the beauty of the place.

I have since lived on both the east and west coasts of the U.S., have spent time along the Gulf Coast, and lived in Hawaii. While I have also lived in places landlocked and far away from the oceans, it seems I can't get myself to leave the ocean for long.

Even as I am writing this book I am taking breaks to bike to the beach and jog in the sand, or am sitting by the beach while I write.

I love the ocean water, the scenery, and being by the shore during the sunset, sunrise, and winter storms and hot summer nights. I like the changes in the surf, and seeing birds, otters, seals, sea lions, dolphin, and the occasional whale. I like running along the beach, swimming, and snorkeling. I often sit on the beach while writing, reading books, talking with friends, or sharing a meal. I've camped on many beaches, attended all-night parties on beaches, and have shared many great times with my favorite people there.

"A lake carries you into recesses of feeling otherwise impenetrable."
– William Wadsworth

Water pollution

One thing that really gets me about beaches is how increasingly polluted they are. Even when I was in the middle of the Pacific, standing on the south shore of the Big Island of Hawaii, at my feet were scattered a variety of plastic bits, some of which likely consisted of trash dumped off boats and ships, and other varieties of plastics that had floated for hundreds – or thousands – of miles before washing on shore.

Plastic pollution is only one of the many lousy things going on with the oceans. As this book details, the oceans – and all life in and around them – are under constant bombardment of pollution, development, and environmental changes that are not good. Humans have built up to the edges of the oceans, blocked rivers and washes that would otherwise flow into the oceans, covered marshes with buildings, roads, and parking lots, built resorts and industrial complexes on the coast, and otherwise have destroyed millions of acres of what had been part of the circle of Nature helping to keep the oceans and wildlife in healthy balance.

Ocean acidification

"The change we are seeing at the moment is taking place extremely rapidly, and it is the speed of change. We are seeing levels of pH now that probably haven't been experienced for 55 million years. We've gone back a long time. The time of the dinosaurs, put it that way. I find it very difficult to tell people what a scary situation we are in at the moment. The oceans are changing in a huge way. And I'm particularly worried for my grandchildren. The changes we thought were happening in the future, but we are actually seeing them happening now. And they will affect all of our lives. The whole of the world in the next few decades to 50 years. We will see great changes taking place in our world."
– Professor Chris Reid, Marine Institute, University of Plymouth and Sir Alister Hardy Foundation for Science; 2011

Puzzle of Nature

As I became more aware of what is going on in and around the oceans and land, I started to piece together what has turned into this book. At first, I put it together into a long essay-like piece. Certain people who had read it encouraged me to publish it as a book to educate people about what is going on with the environment and wildlife.

Earth's immune system

What humans have done to Nature is punching us back in ways directly impacting the ability of humanity to survive. The storms, heat, and crop failures are like the Earth's immune system trying to shake off humanity. Unfortunately, the defensive measures are also scarring the planet, altering its function, crippling it, and shortening its lifespan as a living thing. Many parts of the circle of life are vanishing as various types of plants and animals are going extinct, damaging the ability of other forms of life to survive.

Extinction rate

"Before the Neolithic period 1,000 years ago, the extinction rate was – as close as we can tell – about one species going extinct per million per year. And the rate at which species were being created before humanity plunked down in the middle of things was the same. So you had very roughly an equilibrium. Even though species were coming and going, it was a very slow turnover. We have now upped the extinction rate by at least 1,000 times. It could easily to go 10,000. We have eliminated so much natural habitat that we now have places all around the world where we've cut everything down to 10 percent or 20 percent of the original cover. That's what you see in the Philippines, in Madagascar, in Hawaii. Substantial parts of Indonesia are going that way fast. Another very rough figure: if you reduce an area by 90 percent, so you have only 10 percent of the original forest left, for example, whatever the habitat is, you can expect to sustain only about half the species. Half of them, roughly, will go extinct. This is what we learned from our theory of island biogeography, and it happens pretty quickly.

We are also reducing the birthrate – the cradles are disappearing. Now, why would I say that if we're not very careful we could easily take it from 1,000 up to 10,000 times the pre-human extinction rate? Because it's so easy to knock out the land that remains. That's why I think we can use those high figures. Some experts might be more conservative, but that's where we're at now. That's why we have to create preserves all around the world, make them as large as possible, and also connect them up with one another. Most scientists working in conservation and biology would agree. We need corridors, particularly in the face of climate change."

– E.O. Wilson, founder of the inquiry of sociobiology, and co-author of *The Theory of Island Biogeography*, and Pulitzer-prize winning co-author of *The Ants*; OnEarth magazine, winter 2011; OnEarth.org

Situation dire as ever

The environmental situation is as dire as ever. I didn't want to write a long, detailed book that would take too long to research and write, or require a big commitment for someone to read. I wanted the information out as soon as possible so people could understand the dire circumstances of the bad choices of humanity. Hopefully, they will become part of the solution.

As I was researching pollution, environmental damage, wildlife endangerment, and extinction, I realized how firmly the whole web of life is connected. Covering all of that could result in a book series tens of thousands of pages in length – and still not cover it all.

Every type of life on the planet is in trouble.

Network of Nature

From the trees and animals of the forests, the plants and animals in the water, the life of the planet is reliant on the network of Nature.

All Earth life is being impacted by environmental degradation.

We could all be helping

We could all be doing something to heighten awareness, to encourage action, to live more sustainably, and to restore Nature.

Perhaps by putting this book out I could influence some change for the better, even so much that I might inspire people to make meaningful changes in their lives and the lives of others. If there is any way we can help sway humanity to change their ways and to be involved in taking care of and restoring Nature, now is the time to do it.

Water

"As man increasingly overcrowds and exploits his tiny planet, the significance of the oceans which cover seven tenths of its surface has suddenly become apparent."
– Margaret Deacon

"A lake is the landscape's most beautiful and expressive feature. It is Earth's eye; looking into which the beholder measures the depth of his own nature."
– Henry David Thoreau

Throughout history, water has represented many things. Water is mentioned in ancient texts, scriptures, myth, lyrics, poetry, parables, novels, and other works. People draw water, paint it, photograph and film it, sing about it, swim and bathe in it.

Water is present in symbolism, in myth, in religious events, in food, and in plants and animals, including the human animal.

From bacteria, mold, microorganisms, the creatures in the water, on the land, and that fly, and all of the varieties of plants, every living thing on this planet relies on water.

Without water constantly being present, we would not survive.

Humans are the only life form damaging Earth

Yet, humanity, the so-called highest form of intelligence on the planet, has done more to damage the water on this planet than anything else.

Humanity is the only life form that has damaged – and is damaging – the water sources, lands, plants, and animals of the planet.

"River flows in Turkey, Syria, Iraq, Lebanon, and Jordan have been depleted by 50- to 90-percent in the last half century. The flow of the Jordan River, a primary water source for five of these countries, has decreased 90 percent since 1960."
  – *Water for Peace*; United Press International; Jan. 1, 2011

Unseasonable and out-of-balance

With rising temperatures, evaporation rates increase, desertification spreads, flowers bloom out of season, birds and other wildlife build their nests, dens, and other nursing structures out-of-season, migration patterns change, and deeper in the soil and oceans, the lives of the very small, including the bacteria and fungi which support other forms of life, grow out-of-balance, out of control, or not at all.

Nature disrupted

In the waters of the planet, warmer temperatures disrupt the mating and growth of many types of fish, crustaceans, and smaller life forms, and that impacts the existence of larger creatures who feed on the smaller ones. This all plays into the nutrients in the oceans the marine plants depend on, and that impacts the amount of oxygen the oceans produce.

Atlantic cod, another example of a species in danger

The Atlantic cod is often given as an example of a fish impacted by the higher water temperatures. Warmer waters increase the activities of the cod, increasing their metabolism, and burning more calories. This results in smaller fish.

The warmer waters also impact the growth of the natural food sources of cod: herring, smaller fish, and crustaceans.

What is going on in the seas is similar to what is going on across the continents.

Fires and flooding and deadening fury

While some regions of Earth are experiencing record heat and draught, leading to fires, other areas are experiencing record rainfall, leading to massive flooding. Each of these scenarios impacts all varieties of wildlife.

In recent years massive fires have erupted in various parts of the world, incinerating villages and towns, forests, farmland, meadows, and scrubland, killing scores of people, and unknown varieties and numbers of animals and plants.

As I'm writing this paragraph in the summer of 2023, there are fires raging in various regions around the globe. At the same time, there are storms causing massive floods. Droughts are also being experienced at record-breaking temperatures with heat waves that keep intensifying.

Deep emergency

"The public simply doesn't understand, in my opinion, what a deep emergency we are in. This is the merest beginning of what we are going to see in coming years. And, to me, it is absolutely horrifying. I don't think people fully appreciate how irreversible these impacts are. We can't reverse this. It's not like cleaning up trash in a park.

How hot we allow this planet to get is how hot it will stay for a very long time.

I feel like climate scientists – including myself – have been being ignored for decades by world leaders. They don't seem to get this, either.

I'm glad to hear President Biden finally using his bully pulpit a little bit to try to wake up people a little bit that this is real. But he continues to expand fossil fuels at breakneck pace. He continues to permit more drilling on public lands at a pace even faster than Trump, to approve the Willow Project in Alaska. He went out of his way to make sure the Mountain Valley Pipeline – a natural gas pipeline in Virginia and West Virginia and was approved. He could have stopped that. But instead he's pushing to expand fossil fuels. And fossil fuels are the cause of all of this damage that we're seeing. The deadly fires in Greece, in Maui, the flooding we've seen in Vermont this year, and in Pakistan last year that practically inundated most of the country.

The record heat we're seeing is going to get worse and worse.

I feel like we're are on the verge. These are very non-linear changes. It feels like they are increasing very quickly, because they interact with society in very complex ways. And we are a lot more vulnerable than most people think, or thought quite recently.

We can start seeing things like regional heat waves that end up killing a million people in the course of a few days in coming years. And it won't stop there. It gets worse with the more fossil fuels we burn.

Doing the science, publishing the papers, hasn't seemed to get the message across – either to the public, or to world leaders.

I've got two sons, and it breaks my heart to see the Biden Administration continue to expand fossil fuels, and take us deeper into this catastrophe – instead of trying to take us back from this. He's deeply on the wrong side of history.

Choosing JP Morgan Chase Bank in Los Angeles [where he held a protest] last year [2022] was a strategic choice, because a lot of these new fossil fuel projects. And let me say again how insane it is we are still allowing new fossil fuel projects to be built – because they have lifetimes of three-to-four decades. The financing of those projects is crucial, and no institution does more irreversible damage to the Earth system by financing fossil fuel projects than JP Morgan.

The public needs to know that the fossil fuel industry and its leaders, and their lobbyists, have been lying for decades. For about 50 years. This is well documented. There is a paper trail.

There is a clear body of evidence that the fossil fuel industry – and through various organizations like the American Petroleum Institute – have been literally lying to the public, trying to spread confusion about the science, countering climate scientists' attempts to sound the alarm, creating this sense of uncertainty through their lies. Spending billions of dollars on these misinformation campaigns, and then bribing politicians.

A year ago a story in the *New York Times* said we all know (West Virginia) senator Joe Manchin received a lot of money from the fossil fuel industry. But even (New York) senator Chuck Schumer received almost $300.000 in one election cycle from the corporation that benefits from the Mountain Valley Pipeline – to ensure the Mountain Valley Pipeline was built.

The tendrils of the fossil fuel industry, and it's surprising how cheap it is to buy off these politicians. It reminds me of the David Bowie song *The Man Who Saved the World*.

I know that with President Biden, during the primaries a lot of the people in his campaign team had worked previously in the fossil fuel industry. So there are a lot of connections there, as well.

Part of the problem is simply that we have one of the most powerful industries on the planet – if not the most powerful industry – which has extremely deep pockets. They have profits of over, I think, a trillion dollars per year. They can spend a tiny bit of that money to basically influence politicians. It's essentially legalized bribery.

Their disinformation campaign is a big part of why the public doesn't understand how serious of an emergency we are in right now. And that in turn kind of doesn't push journalists to connect these dots.

I see a lot of stories being reported in *The New York Times*, and elsewhere about these individual climate catastrophes, but they miss key points in the story. First of all, they often use the passive voice. They say, 'Oh, like the Earth is heating up." No, it IS being heated up by the fossil fuel industry, by their dishonesty, by their legalized bribery.

So they [the journalists] don't make that connection. They also don't make the connection of where we are going in the near future. So, if they are talking about a deadly heat wave that happens in 2023, they don't say how much worse things are going to get by 2028, or 2032.

This is what really frightens me about climate change and global warming caused by global heating, it's a trend. You might have some years that are naturally cooler than others due to natural variability. So there is a little bit of a noisy trend. But it is rising year on year. The physics is absolute. You can't negotiate with it. We understand the physics quite well. We don't understand how it is going to play out with these complex human systems, like the agriculture system, the water systems, geopolitics – that's a whole other question. But we know it is going to get hotter and hotter, and that's going to drive all of these types of catastrophes that we are seeing to get more intense, more frequently."
– Peter Kalmus, author, NASA climate scientist

Fires and radiation

In the past dozen years fires have burned across landscapes contaminated with radiation left over from government testing in New Mexico and a nuclear accident that happened decades ago near Simi Valley, California. This magnifies the complexities of fire pollution.

Diet of destruction

How many acres of land are humans going to ruin before the majority of people finally notices the problem? But they keep participating in it, mostly with their mouths – by what they eat.

Destroying Nature for meat

The massive industrialization of the cattle industry, including monocropping millions of acres of land across every continent to grow feed for billions of farm animals, has resulted in degraded land trampled and compacted by cattle and huge machinery. Land used in this way becomes devoid of diverse plant matter that would otherwise support a variety of wild animals. Millions of wild animals have been killed specifically to protect chemically-grown feed crops.

Killing predator animals to raise livestock

Here, in North America, a few hundred centuries of killing predator animals to protect farm animals has brought about more deer living in areas where they were formerly sparse. This has caused the loss of certain types of plants because the out-of-control deer populations are consuming more plants than ever. This strips the land of even more bushes, trees, and smaller plants that build soil, contribute to soil health, sequester carbon, build root systems that prevent floods and wind damage, and plants of all varieties that are both food and homes for birds, bees, butterflies, bugs, and other forms of wildlife.

Soil compaction and loss

When the rains fall on land compacted by cattle and stripped of its natural plant matter, monsoon flooding occurs. Instead of soaking into soft soil and into aquifers, the rains form into rushing streams that build into wide floods that wash away soil that took unknown numbers of years and combinations of Nature to build, but which gets carried away into rivers, suffocating fish and other waterlife.

Of livestock and feed crops

As floods spread across farmland and into farming communities, it has exposed some of the damage humans have done by trying to control and redirect rivers by building dams and flood control channels; by turning wetlands into fields of feed crops to support the cattle industry, and by altering landscapes to build highways and towns. It is the flood planes that would have naturally absorbed the water, allowing it to seep into the soil and replenish underground aquifers. Instead, the water is thought of as a nuisance when it ends up where it naturally spread across the ancient flood plains now altered by the cattle, dairy, feed crop farming industries, and human sprawl. In America's Midwest, instead of seeping into the land, and into aquifers, the storm water ends up in rivers, and in the Gulf of Mexico, bringing with it millions of tons of sediment saturated with farming chemicals.

Chemicals used on livestock feed

As more chemicals are being used to grow enormous quantities of food to feed hundreds of millions of farmed animals, and more drugs are being used on the farm animals, the chemicals and drugs also end up in the aquifers, ponds, wetlands, rivers, lakes, marshes, and oceans – and in drinking water, in the body tissues of wildlife, and in us.

Hormone-disruptors

Hormone-disrupting chemicals are an increasing problem around the world. As the use of plastics; cleaning and personal care products containing chemicals; pharmaceutical drugs and their vast array of chemicals; insecticides, pesticides, and other farming chemicals – such as atrazine, which is used on corn fields; and other sources of chemicals expands, more and more problems are occurring among wildlife, such as aquatic creatures being found with physical defects, skin lesions, cancers, extra limbs, and with the organs of both sexes.

Damage by livestock and feed crops

The floods happening in many countries each stand as examples of what the soil compaction, deforestation, urban sprawl, and the overgrazing and land and wildlife degradation caused by the cattle and feed crop industries are doing to the planet.

Biodiversity loss

"The use of land for agriculture is the main driver of biodiversity loss. Today, almost half of the world's ice- and desert-free land is used for agriculture, and most of this land is used by livestock. The total global land used for meat and dairy production sums up to 37 million square kilometers, an area as large as the entirety of the Americas – from Alaska in the north to Cape Horn in the South."
  – Max Roser, *How many animals get slaughtered every day?*; OurWorldInData.org, Sept. 2023

Australia cattle industry, cotton, soil, and storms

As on other continents, large areas of Australia have been impacted by both cattle grazing and the growing of crops to feed to farmed animals. (Australia also imports animal feed.)

The cotton industry is one that also has damaged the land in Australia. It is a place where cotton, which is a water-intensive crop, shouldn't be grown – at least not on the scale it is being grown. (Hemp would be a better choice as it takes less water, is better for the land, sequesters more greenhouse gasses, produces more varieties of fabric than cotton, and more varieties of materials – including seed for food and oil, and other materials from the hemp plant to produce resin, ink, fiberboard, flooring, insulation, hempcrete, air filters, cellulosic ethanol, compostable plastic, etc.)

Recent years have been a time of devastation for Australia. Intense rainfall has caused massive floods that ruined farmland, neighborhoods, and roads, and engulfed parts of cities. Crops have been destroyed, and mass quantities of sediment containing farming chemicals have washed into the ocean where Earth's largest living organism, the Great Barrier Reef, sits. The pesticide- and fertilizer-saturated sediment kills microorganisms, increases algae growth, bleaches the coral, and causes the death of many type of sea life dependent on the reef. Massive fires in Australia have burned many homes and businesses, incinerated forests, and driven koala bears closer to extinction. As I write this paragraph in December 2023, massively ruinous floods are happening in Australia.

Planetary environmental hopscotch

We can skip around the continents for examples of what humans have done to Earth in ways that places humanity – and all life forms on the planet – at risk. Including by damaging and decreasing food sources.

Vanishing Aral Sea

One of the awful examples of what humans have done to damage a region of Earth is what they have done to the Aral Sea.

The Aral Sea was once one of the largest bodies of fresh water on the planet. From the 1950s to the 2010, over 75% of the Aral Sea had been drained. Much of it was used to irrigate unsustainable cotton farms. While the Aral Sea was once easy to spot on continental maps, at its current size, it has become a small speck compared to what it once was.

"It is a curious situation that the sea, from which life first arose, should now be threatened by the activities of one form of that life."
– Rachel Carson, author of *Silent Spring*

Nuclear disasters, waste, and radiation pollution

Along with the Chernobyl nuclear power plant meltdown in Ukraine, the Fukushima nuclear power plant disaster of 2011 continues to be a widespread nuclear disaster. Since the accident, radiation is being found in Japan's tap water, and in the ocean off the coast of Japan, and in the life within the water. Tuna and other migrating marine life throughout the Pacific have been found containing radionuclides from Fukushima.

Scientists say the radiation from Fukushima will dissipate until it is harmless. Or, will it. You can find out more about it for yourself as science journals online continue to follow the complexities of the accident – including what the plant is doing with the millions of gallons of radioactive water it has been accumulating in storage tanks.

The accident brought about a reliance on bottled water. Millions of bottles of water have been sold because people have become afraid of drinking the tap water. Along with the bottled water comes more plastic and packaging, which ends up as trash, and a whole lot of it ends up as trash in the water.

Radioactivity water pollution

As I'm writing this paragraph in August 2023, it is twelve years after the Fukushima nuclear power plant disaster. The radioactive reactor cores continue to contaminate the water pumped into the plant, as well as rainwater that seeps in, and ground and ocean water. Japan is planning on releasing millions of gallons of "treated radioactive water" into the ocean. They have built over 1,000 tanks holding nearly 1.4 million metric tons of radioactive water. They say space has run out, and they have no more room to store more radioactive water, which contains a hydrogen isotope called radioactive tritium. With no technology invented to remove the isotope from the water, it will be poured into the ocean.

While radioactive tritium exists naturally in humans, other animals, and the environment, releasing the water will greatly magnify the amount in the ocean, and in marine life – including sea creatures sold as food for humans.

China has banned fish imports from Japan. Is it an overreaction? Do your own research, and learn.

American standards for safety of radioactivity in water is 740 becquerel per liter. The World Health Organization safety measurement is 10,000 becquerels per liter. Japan's standard is much higher, at 60,000 becquerels per liter. What is safe? Who knows. It's a guess.

Tritium pollution

Here is something you probably didn't know: The hundreds of nuclear power plants around the planet routinely dump water containing tritium into rivers, lakes, and oceans. Research it.

Radiation bioaccumulation in fish

Scientists debate the potential health risks on humans caused by the release of Fukushima's radioactive water, including the bioaccumulation of it in plants, animals, and bacteria, and how it will impact the food chain of wildlife and humans. Protests held during the summer of 2023 in Japan and Korea citing the unknown risks of dumping the radioactive water into the ocean. People have been stocking up on canned fish and sea salt. The Japanese fishing industry has been negatively impacted since the nuclear plant was damaged.

The research hasn't been conducted to conclude one way or another of how damaging or safe continued dumps of tritium are to wildlife. One animal or plant might react differently than others.

Tuna off coast of California found with radiation from Fukushima

People know I've been involved with environmental issues. Some have asked me about the news stories of radiation in tuna, as if the news story can't possibly be correct.

You can research this for yourself, and find more detailed reports.

August 2013: Hearing of tuna caught off of California and Mexico contain radiation from Japan's nuclear power accident should not be a surprise.

Tuna caught off the coast of the U.S. have been found to contain radiation traced to the ongoing nuclear disaster in Japan. (Search: Fukushima nuclear accident contaminating marine life.)

As radiation continues to spread into the ocean from that horribly-designed and badly damaged nuclear power plant, the more the sea creatures will bioaccumulate the radiation. The small creatures get eaten by the bigger creatures, and each step gathers more radiation – and other industrial pollutants – on up to the larger creatures, like tuna, turtles, dolphin, seals, otters, and whales. Then into marine birds and into land animals that eat sea creatures along the coast, and from rivers. These include bears, foxes, cougars, bobcats, and raptors.

Listening to the FDA as your advisor to determine what you eat might not be what is best for you. Do your own research on the topic.

The FDA might be considered a spokesmodel for the corporate interests, including the ruling corporations in the fishing industry.

If you consider its history with how it has given into pressure from the fishing industry, you might conclude that the FDA does not act in the best interests of consumers.

The FDA might be viewed as a concierge service for industry. Listening to them for advice on which foods to consume is like listening to them on which drugs are safe. The FDA has a horrible record of approving drugs that then cause deaths, miscarriages, nerve damage, cancer, birth defects, strokes, blood clots, infections, kidney failure, blindness, tissue necrosis, loss of smell and taste, etc.

This isn't to say I am anti-vax. I did get the COVID vaccine, and got boosted.

The FDA, the USDA, the Nuclear Regulatory Commission, the Department of the Interior, the State Department, the Department of Justice, the Pentagon, etc... ALL work in the interests of corporate profits. Be it the fossil fuels industries, the prison industry, the nuclear energy industry, the farming chemicals industry, the plastics industry, the animal agriculture industry, the fishing industry, etc. Many of the industry leaders have worked for the FDA, and/or go to work for the government.

Consider what each of those industries has done to the environment and human health, and you might conclude that every department of the government has been corrupted by corporate interests.

Public uproar successfully gets nuclear power plant shut down

If it weren't for public uproar, the nuclear energy plant in Orange County, California would still be operating. The National Regulatory Commission would have allowed it to continue operating with all of its dangerous flaws. Because public uproar exposed the flaws and dangers of the nuclear power plant, it isn't generating power, but it remains an enormous problem. The nuclear waste it generated during its decades of operation remains stored on site. A major earthquake, a plane crash, a bomb, an under-ocean landslide, or a number of other events can cause that nuclear power plant – and the other nuclear plant up the coast, Diablo Canyon Nuclear Power Plant – to release tremendous amounts of radiation that would quickly contaminate some of the most dependable farmland on the continent, and force the permanent evacuation of millions of people.

Where would you move millions of people from Orange County, as was done to the people in the region of Chernobyl in the 1980s?

Chernobyl nuclear meltdown aftermath

For some interesting videos, go to YouTube and look up Chernobyl. The city sits abandoned, rotting away. There are trees growing on rooftops, balconies of apartment buildings, and office towers. Many surrounding towns sit abandoned, trees and bushes growing through rotting houses. Vast amounts of farmland had to be abandoned.

Consider that there are dozens of nuclear power plants in the U.S., and all are potential enormous problems, such as the one near New York City. If that nuclear power plant had a meltdown or a burn-through, where would we move everyone in New York City and the surrounding region, permanently? What would it do to the world economy?

Dust, smoke clouds, and weather spread radiation into atmosphere

Within weeks of the nuclear accident in Japan, the radiation from it spread through the atmosphere.

If you consider what happens when there is an enormous dust storm in Africa, or there is a large volcanic eruption, and understand how fast the dust clouds or smoke spread into Earth's atmosphere. You get an idea of what happens when a nuclear accident happens.

In 1991 the Mount Pinatubo volcanic eruption in the Philippines spread gasses and solids from the eruption into the stratosphere, and that circled the globe within three weeks.

Since the damage to the Fukushima nuclear power plant caused by the combination of the earthquake and tsunami, the radiation has been spreading through the atmosphere, clouds, rain, oceans, and wildlife food chains. As Japan releases more of the radioactive water, the radiation continues to spread.

Chernobyl nuclear accident

Farm animals in Europe are still being found with radiation from the Chernobyl nuclear accident, which spread radiation across Europe's most used farmland. The farm and wild animals eat the food and plants grown in Europe, and bioaccumulate the radiation.

Consider how many people are still getting cancer directly linked to the Chernobyl accident. How many birth defects are still happening because of it. How many institutions are filled with children and now adults there with birth defects caused by Chernobyl.

I haven't eaten animals, dairy, or eggs in over three decades. Perhaps I've bioaccumulated less than average amounts of radiation.

Animals bioaccumulate environmental toxins

Animals collect toxins from their environment, including farming chemicals, industrial pollutants, pharmaceutical drugs, and radiation. When you eat meat, milk, or eggs, you eat some of whatever toxins those animals accumulated.

It is healthful for humans to be eating low on the food chain: This means to eat plants = fruits, vegetables, sprouts, nuts, and seeds. Especially organically grown in your region.

Nuclear reactors

As the news is continually filled with more and more foolish nonsense, people are more disconnected from important things that impact them. They are limiting themselves to hearing about and obsessing over so-called "news" stories that have absolutely no impact on their lives – including celebrity culture - and all of that other garbage in the news that doesn't matter.

The popular American news sources are a sick joke. It appears they don't know real journalism, and seem only concerned about how they appear on camera as they spew corporate messages.

Look up environmental and wildlife protection organizations. Consider what impacts your foods, air, water, and life.

There will be more nuclear power plant meltdowns

"Apollo 13, as you may remember, gave us a reactor that is bubbling away right now somewhere in the Pacific. It's supposed to be bubbling away on the moon, but it's in the Pacific Ocean instead."
– David R. Brower

Fukushima and Chernobyl won't be the only nuclear power plant disasters. Each nuclear power plant is a potential disaster. Many sit on the edge of oceans, and are susceptible to storms, earthquakes, tsunamis, and other natural and man-made disasters, such as an airplane crash, bomb, or the climate breakdown caused by human activity.

Dirty nuclear power plants

Nuclear energy is filthy, destructive, costly, and dangerous. This will especially be so if the electric systems end, and there will be no way of preventing the current 437 global nuclear plants from melting down, and experiencing explosions. The Fukushima and Chernobyl nuclear plant explosions spewed radionuclides into Earth's atmosphere, which spread across the land, and into rivers, lakes, marshes, and oceans.

In 2023, a new nuclear power plant opened in Georgia.

Building a nuclear power plant requires an enormous amount of money, fuel, concrete, and steel, and a wide variety of machinery to build, maintain, and run the facility. All of that equipment needs to be sourced from the planet, including the steel, concrete, and fuel.

Uranium mining

Nuclear plants use uranium, which is mined, leaving behind gaping radioactive wounds in Earth. The mining areas remain toxic for at least centuries. Many of the U.S. mines are on Native American reservations, and have increased the rates of cancer among the same people who have been mistreated and violated for centuries.

Nuclear waste

A nuclear power plant produces nuclear waste. This means there needs to be a place to store the waste, which remains hazardous for thousands of years. There are now thousands of sites holding nuclear waste, and more are being created – resulting in more and more parts of the planet dangerous to humans and wildlife, and problematic for many centuries to come.

Susceptible nuclear power plants

All nuclear power plants are susceptible to natural and man-made disasters, and to radioactive meltdowns. As mentioned, they then spread radiation throughout both the surrounding and distant environment. The radiation then bioaccumulates in wildlife, and humans.

"Plankton absorbs the cesium, the fish eat the plankton, the bigger fish eat smaller fish – so every step you go up the food chain, the concentration of cesium gets higher."
– Michael Friedlander, a U.S.-based nuclear engineer, March 31, 2011. Speaking of the radiation being found in seawater off the coast of Japan.

Long-lasting pollution

Every time humans dig into Earth to extract minerals, coal, oil, natural gas, ore, plutonium, and other substances, more toxic waste is created, more water is polluted, and more wildlife is killed and/or left subject to the dangers of the leftover toxins.

Gold mines

In South Africa gold mines contain water poisoned with a variety of substances used in the mining processes. There are mines filling with water and creating problems with the drinking and farm irrigation water sources, and polluting rivers and lakes. This will be an ongoing problem for people of the region, including in Johannesburg.

Water is not "away"

With water, humans seem to think, "Out of sight, out of mind." For centuries, humans have dumped unwanted things into lakes, rivers, and oceans. Apparently, humans think oceans, lakes, and rivers are the "away" in the phrase, "throw it away."

Rivers, lakes, and oceans are where humans have dumped everything from murdered people to demolished buildings to massive piles of trash to stacks of bombs and guns to barrels of chemical waste, and a variety of other hazardous materials. Even religions have the teaching that baptism will "wash away your sins," as if we can even dump our so-called spiritual flaws into water. What humans have dumped into the waters of the planet is the real sin.

Marine life extinction

"I think that if we continue on the current trajectory we are looking at a mass extinction of marine species. Even if only coral reef ecosystems go down, which it looks like they will, certainly by the end of the century, that in my mind would constitute a mass extinction event. And that is simply because of the large diversity associated with those systems. In terms of the actual reef-building corals themselves, and also the myriad of organisms.

There are some scientific estimates; there are something like up to 9 million species associated just with coral reefs. However, many of the symptoms, if you like, that we are seeing of change in the oceans indicate that the effects will be much wider than just coral reef ecosystems.

Rising temperatures are already changing distribution of organisms. Animals are moving further north or south basically attracted to the temperatures they like to live in. And, of course, the animals living in the tropics and at the polar extremes of the oceans, they don't have anywhere to go. So there is a good chance we will see high levels of extinction both in low latitudes and in high latitudes, simply as a result of temperature change."
– Dr. Alex Rogers, Professor of Conservation Biology, University of Oxford; 2011

From the rainforests in the middle of the continents to the deepest parts of the oceans, man's polluting ways are evident.

No separation

There is no separation between the pollution, destruction, and health of the oceans, lakes, rivers, marshes, wetlands, mud flats, grasslands, meadows, forests, mountains, and air. Similarly, there is no separation between the health of the rainforests and of the coral reefs.

Rainforest destruction

"Farmers clearing the jungle to make way for agriculture is the main driver of deforestation in the Amazon. About 7,300 square miles of Brazilian forest were lost every year between 1998 and 2007."
– Reuters, June, 2010

As you read this, rainforest destruction continues. Mostly to supply markets, restaurants, and fast food joints with cheap meat. The negative impacts are global, from land, river, lake, and ocean damage, to air pollution, to global warming, floods, and drought, and extinction.

"Extensive cattle ranching is the number one culprit of deforestation in virtually every Amazon country. The deforestation caused by cattle ranching is responsible for the release of 340 million tons of carbon to the atmosphere every year."
– The Hidden Cost of Burgers; World Wildlife Fund; WWF.Panda.org

River pollution

The number one reason the rivers of North America are so polluted is because of all of the chemicals used to grow food to feed farmed animals, and from the pollution caused by the urine and feces of massively bred animals, and the loss of topsoil caused by them, and from growing millions of acres of monocropped fields of feed crops. Then there is fossil fuel pollution, industrial pollution, plastic pollution, pharmaceutical pollution, and technology pollution.

Feed crops

More than 50% of the food grown on the planet is grown to feed farmed animals. More than 85% of some crops go to feed farmed animals. This means more than half of the fuel, water, and land used for growing food goes to support the animal farming industry. A vegan diet does not support these wasteful and Earth-damaging practices.

The great waste of resources used to support the animal farming industry, including cattle, chickens, turkeys, lambs, hogs, and other animals, is supported by those who consume meat, dairy, and eggs.

Five key measures

"We studied 55,000 people's dietary data, and linked what they ate or drank to five key measures: greenhouse gas emissions, land use, water use, water pollution, and biodiversity loss.

Our results found vegans have just 30% of the dietary environmental impact of high-meat eaters.

Vegans in our study had just 25% of the dietary impact of high-meat eaters in terms of greenhouse gas emissions, for instance. That's because meat used more land, which means more deforestation, and less carbon stored in trees. It uses lots of fertilizer (usually produced from fossil fuels) to feed the plants that feed the animals. And because cows and other animals directly emit gas themselves."

– Michael Clark and Keren Papier, *Oxford Scientists Confirm Vegan Diet is Massively Better for Planet*; ScienceAlert.com, July 24, 2023

Consider meat

Those who consume meat may want to take another environmental factor of their diet into consideration, this is the fact such a large meat and dairy industry could not exist if it were not for a tremendous reliance on the use of fossil fuels, nonstop production of plastics, and the massive use of refrigeration, and at the expense of other forms of life, including fish, sea mammals, and marine birds, and life dependent on these animals.

Fish fed to livestock

"Forty percent of the fish taken from the oceans is fed to livestock. This means pigs and chickens are becoming major aquatic predators.

The livestock industry is one of the greatest contributors to greenhouse gas emissions, ever.

The eating of meat is an ecological disaster."

– Paul Watson, Sea Shepherd Conservation Society founder, Autumn 2010

Reliance on corporate foods

Those who are completely dependent on restaurants and stores for every calorie may want to consider what their reliance on corporate foods – and all of the pollution it causes – does to the environment and wildlife. This is even more if they are consuming meat, dairy, and eggs – which lead to an enormous amount of packaging, plastics, fossil fuel use, water pollution, land and forest degradation, and greenhouse gasses.

Main use of plastic

With the so-called modernization of the food industry came the use of plastic. The number one use of plastic on the planet is now related to food packaging, preparation, and service. At one time or another, the plastics end up as trash filling landfills, scattered across the landscape, and in the streams, ponds, rivers, lakes, marshes, and seas.

Marine life dying from plastic

Millions of fish, sea mammals, marine birds, and some land animals end up eating so much plastic they eventually die with digestive tracts blocked with the variety of plastics they have swallowed.

As mentioned earlier, the impact of plastic pollution isn't only happening in and around cities and towns, but is happening on distant islands and in the most remote places of the oceans, all the way down to the coral reefs and ocean floor.

Plastic and bird deaths

Plastic floats. A tremendous amount of plastic is floating in the oceans, including in parts of the oceans as far as you can get from land.

It is becoming increasingly common for dead seabirds to be found with their guts filled with a variety of plastic trash, and for fishing companies to find plastic inside the fish they capture in distant seas.

Humans: a destroyer

"Man has been endowed with reason, with the power to create, so that he can add to what he's been given. But up to now he hasn't been a creator, only a destroyer. Forests keep disappearing, rivers dry up, wild life becomes extinct, the climate is ruined, and the land grows poorer and uglier every day."
– Anton Chekhov, *Uncle Vanya*, 1897

Earth, the trash dump

Every toxic substance humanity has ever created can be found in rivers, lakes, and oceans, and in the underground aquifers of the continents. And the practice continues. Even as you read this there are people someplace on earth dumping trash, chemicals, cars, plastics, industrial waste, military garbage, cruise ship trash, and every toxic thing into some area of the planet's oceans and lakes and rivers. Humans also use canyons, caves, old mines, and wells as trash receptors.

Mercury pollution

"Give a man a fish, and he can eat for a day. But teach a man how to fish, and he'll be dead of mercury poisoning inside of three years."
– Charles Haas

The fossil fuels being burned day and night all over the world spew toxins into the air. These toxins – including mercury from growing numbers of coal-fueled electric generating plants and concrete kilns – add to the problem of polluted waters.

Oceans, lakes, and rivers naturally absorb air pollution, which helps lead to acidic (sick) rivers, lakes, and oceans, which poison and kill waterlife, and leads to toxins being in the tissues of the animals feeding upon the fish, sea mammals, and marine birds.

Water dance

"Water, like religion and ideology, has the power to move millions of people. Since the very birth of human civilization, people have moved to settle close to it. People move when there is too little of it. People move when there is too much of it. People journey down it. People write, sing, and dance about it. People fight over it. And all people, everywhere and every day, need it."
– Mikhail Gorbachev

We all need clean water for wildlife to thrive, and to grow our food, hydrate our bodies, and refresh us.

Altered water life

Much of the world's water is so polluted the life within it is altered, is changing, is moving to different locations where they also struggle to survive, or is dying.

Aquifers, ponds, lakes, rivers, and other sources of water are being depleted by supplying water for golf courses, for lawns, for water-hogging toilets and dish washers, and for farms growing food for the world's billions of farmed animals (including fish farms, animal factory farms, and cattle ranches).

Each one of us can change things in our lives to protect the water of Earth, which is really what we are. What ends up in the aquifers, rivers, lakes, marshes, and oceans ends up in the wide variety of life on this planet, including in us.

"There is but one ocean though its coves have many names; a single sea of atmosphere with no coves at all; the miracle of soil, alive and giving life, lying thin on the only Earth, for which there is no spare."
– David Brower, first executive director of the Sierra Club; founder of Friends of the Earth; founder of Earth Island Institute; considered the father of the modern environmental movement

Greener products

Please protect water by using not only biodegradable, but compostable, plant-based cleaning products. Also, reduce your use of plastics and fossil fuels, grow some of your own food using organic techniques, compost food scraps, support local organic farmers, follow a plant-based diet, use water-conserving faucets, use non-toxic paints and finishes, and eliminate the use of toxins in your household – including in personal care items, medicines, and cleaning products. Shop less.

Ecosystem collapse

"A friend of mine who is a scientist at the Seattle Aquarium says that she and her colleagues believe the oceans are on the verge of complete ecosystem collapse. And nobody knows about it. None of my friends, even well informed friends know about this. And I think it's because it's all under the surface, literally under the surface where it can't be seen."
– Peter Heller, author of The Whale Warriors

"Not energy depletion, economic collapse, limited nuclear war, or conquest by a totalitarian government. As terrible as these catastrophes would be for us, they can be repaired within a few generations. The one process ongoing in the 1980s that will take millions of years to correct

61

is the loss of genetic and species diversity by the destruction of natural habitats. This is the folly our descendants are least likely to forgive us."
— E.O. Wilson, founder of the inquiry of sociobiology, and co-author of *The Theory of Island Biogeography*, and Pulitzer-prize winning co-author of *The Ants*

"While we've been saying for decades now that this is what to expect, it's still very confronting to see these climate extremes play out with such ferocity and with such global reach.

It's bewildering to see new fossil fuel extraction projects still getting the go-ahead here in Australia. And with this comes deep resentment for those who have lobbied for ongoing fossil fuel use, despite clear climate physics that have been known for almost half a century."
— Professor Matthew England, professor, Australian Center for Excellence in Antarctic Science, University of New South Wales, 2023

"It is distressing to see the widespread damage caused by the current outbreak of extreme events in many parts of the globe. Unfortunately, they are not a one-off, but part of a longer trend fueled by human greenhouse gas emissions. So they are not unexpected.

Worryingly, it is clear that future extremes will again break records, and cause even greater damage. In particular, this is because the damage in many cases is nonlinear – it rises more and more quickly for each increment of climate change. This should cause us concern. It rationally should cause us to step back to assess what is in our economic, social, and environmental interests.

It is also in our interest to put in place large and integrated programs for climate adaptation to deal with the climate change impacts we can't avoid. Taking action to reduce emissions and to adapt to climate change will give us hope. Do we really want the alternative?"
— Professor Mark Howden, director of the Institute for Climate, Energy, and Disaster Solutions at the Australian National University, 2023

"We knew by the mid-1990s that lurking in the tails of our climate model projections were monsters: monstrous heat waves, catastrophic extreme rainfall and floods, subcontinental-scale wildfires, rapid ice sheet collapse rising sea level of meters within a century. We knew – just as we know gravity – that Australia's Great Barrier Reef could be one of the earliest victims of uncontained global warming.

But as today's monstrous, deadly heat waves overtake large parts of Asia, Europe, and North America with temperatures the likes of which we have never experienced, we find even 1.2C of global warming isn't safe."
— Bill Hare, physicist and climate scientist, and CEO of Climate Analytics

The Death of Earth Life

"Think of fisheries without fishes, logging without trees, tourism without coral reefs or other wildlife, crops without pollinators. Imagine the damage to our economies and societies if they were lost. All the plants and animals that make up Earth's amazing wildlife have a specific role and contribute to essentials like food, medicine, oxygen, pure water, crop pollination, carbon storage, and soil fertilization. Economies are utterly dependent on species diversity. We need them all, in large numbers. We quite literally cannot afford to lose them."
    – Jean-Christophe Vié, Deputy Head of International Union for Conservation of Nature's Species Programme; IUCN.org

Links in the chain

Every type of sea mammal, and nearly every type of fish, crustacean, mollusk, amphibian, and reptile is on the critical list. So are seabirds and bears dependent on fish for their survival. Even the microscopic animals of the seas are in trouble. This is because air and industrial pollution is changing the acidity level of the oceans and poisoning sea life; because synthetic farming and lawn chemicals and farmed animal waste are causing massive algal blooms that block light, then rot, lower water oxygen levels, and cause water dead zones; because of fishing, recreational, cruise line, industrial, and military watercraft; because plastic trash is both killing marine life and gathering in and leaching chemicals into the rivers, lakes, and oceans; and because of the burning of fossil fuels, and methane from farmed animals all over the world, and methane from melting permafrost all raising atmospheric temperatures, and the temperatures of the land and water, and melting icescapes. To

63

put it simply, what is going on in the oceans threatens every form of sea life, every form of life dependent on that life, and every human. (Search: Ocean collapse, ocean acidification, and sea life extinction.)

Anthropocene mass extinction

"The synergistic effects of human impacts are laying the groundwork for a comparably great Anthropocene mass extinction in the oceans, with unknown ecological and evolutionary consequences."
– Jeremy Jackson, Scripps Institution of Oceanography at the University of California, San Diego; *Proceedings of the National Academy of Sciences*; 2008

Air pollution

"One of the first laws against air pollution came in 1300 when King Edward I decreed the death penalty for burning of coal. At least one execution for that offense is recorded. But economics triumphed over health considerations, and air pollution became an appalling problem in England."
– Glenn T. Seaborg, Atomic Energy Commission chairman, speech, Argonne National Laboratory, 1969

Since the middle of the 1800s, the amount of carbon dioxide (CO_2) in the atmosphere has increased in relation to the use of fossil fuels (coal, petroleum, and natural gas). The burning of fossil fuels releases carbon dioxide into the atmosphere. Plants naturally absorb and sequester carbon dioxide. But the amount of carbon dioxide being produced by humans is far beyond the amount that could be absorbed by the plants on Earth. It's heavy, and it lands in soil and water. The oceans, lakes, ponds, wetlands, marshes, and rivers also absorb carbon dioxide. But the world's bodies of water are absorbing far more carbon dioxide than they would in a balanced atmosphere.

Ocean acidification

"The oceans matter to us, because we are all connected to the ocean. The oceans are called the blue heart of the planet, and that's very true. Every sip of water you take, every breath you take links you in some way back to how the oceans support us in our day-to-day life. For many people it is really unseen, and they don't worry about the plight of the ocean. But when you understand some of the more direct impacts to employment, to tourism, and then you see some of these less direct impacts of how the oceans supports us, it really is something that should matter to all of us."
– Dan Laffoley, Senior Advisor at the International Union for the Conservation of Nature; June 2011

The industrial pollution and carbon dioxide from the use of fossil fuels have greatly increased and are increasing the acidity of the oceans.

"If you were to ask me, 'what do you worry about the most, what frightens you?' For me, it's ocean acidification."
– Rob Dunbar, oceanographer, biologist

Worst acid trip

Scientists have said the oceans of the world are experiencing the worst acid trips, ever.

Because pollution can hang in the atmosphere for decades, the oceans keep absorbing more of it, and humans keep creating more of it, there are no signs the acid trip of the seas is going to come down soon.

"One of the newest dangers from global warming that has come to light is ocean acidification. As the Earth's seas absorb excess carbon dioxide from the atmosphere, they are becoming more acidic in the process. This changing ocean chemistry threatens tiny, shelled animals, such as amphipods, tubeworms, and crab larva, which are vital links in the ocean food chain. In the Tongass National Forest – the heart of the Alaskan rainforest – pink salmon are losing weight as acidic waters kill off the microscopic pteropods that make up half their diet. The gray whales born in Laguna San Ignacio eat thousands of pounds of these creatures every day in their summer feeding grounds off Alaska. If their food supply crashes, the whale population could follow.

In Costa Rica, a new NRDC BioGem, endangered leatherback turtles are facing the same problem, as more acidic waters weaken the food chain on which the turtles depend. Other climate-related changes post threats to leatherbacks. The temperature of the sand in which they lay their eggs determines the sex of the turtle hatchlings; warmer sands tend to produce more females than males, creating a reproductive imbalance in the population. Rising sea levels can erode and alter the beaches where turtles lay their eggs year after year, further threatening their ability to reproduce."
– Natural Resources Defense Council, Campaign Update; Jan.–Feb. 2010

The increasingly acidic situation taking place in the water bodies of Earth doesn't damage the marine life only at the surface, but impacts marine life all the way down to the bottoms of the swamps, rivers, lakes, marshes, and seas.

Dire situation of coral reefs

One of the most dire situations caused by pollution and the increasing acidification of oceans exists in and around coral reefs throughout the world.

"Ocean acidification is one of the three things we know that has been associated with mass extinctions in the past. We are only just getting to grips with the effects of ocean acidification on marine organisms. The more we learn, the more concerned we become. We are discovering that, not only does acidification of the oceans affect animals that secrete calcium carbonate to build shells, but it also affects the ability of animals to perceive their environment. It actually changes their chemical perceptions through smell of the environment. It affects development, and all sorts of other aspects of physiology of marine organisms. It's really caught the scientific community by surprise. We still do not understand the full implications of it."
– Dr. Alex Rogers, Scientific Director of IPSO and Professor of Conservation Biology at the Department of Zoology, University of Oxford

Coral reefs are among the most endangered forms of marine life. In the web of that life, coral reefs support all forms of ocean life. 24% of all marine life lives directly in and around coral reefs.

Coral reefs throughout the world are on the critical list. Scientists have estimated one-fifth of the coral reefs around the planet have died in the last 50 years, and their tilt toward extinction continues to increase.

Hottest ocean water ever recorded

As I'm writing this paragraph in the summer of 2023, the water off the coast of Florida has measured hotter than any ocean water has been measured on Earth in records kept since the 1800s. Coral reefs there are dying, as is the marine life dependent on them.

"When one tugs at a single thing in Nature, he finds it attached to the rest of the world."
– John Muir

Global Marine Species Assessment

In July of 2008, scientists from around the world gathered at the International Coral Reef Symposium in Ft. Lauderdale, Florida. While there, they were presented with a study by the Global Marine Species Assessment concluding a third of Earth's remaining coral reefs were threatened.

Scientists at the Hawaii Institute of Marine Biology have determined all varieties of coral reefs are the most endangered life forms on Earth.

"Ocean acidification is yet another red flag being raised, carrying planetary health warnings about the uncontrolled growth in greenhouse gas emissions. It is a new and emerging piece in the scientific jigsaw puzzle, but one that is triggering rising concern."
– Achim Steiner, United Nations Environmental Program executive director; Dec. 2010

Coral reefs dying

Thousands of miles of coral reefs filled with life just a decade ago sit almost or completely empty of life because of bleaching, dynamite fishing, overfishing, and rising temperatures and levels of water acidity, or they are being strangled by algae and bacteria overgrowth caused by farming, golf course, landscape, and industrial pollution. An increasing number of coral reefs are simply collapsing.

"The earth we abuse and the living things we kill will, in the end, take their revenge; for in exploiting their presence we are diminishing our future."
– Marya Mannes, *More in Anger*, 1958

Many reefs are being killed by the runoff of synthetic chemical fertilizers used on farms, and the use of gardening chemicals used to maintain the landscaping of tropical resorts; the turf of golf courses and sporting fields; lawns of residences, and the greens of schools, churches, and corporate campuses.

Mississippi River and Gulf of Mexico

When hearing about this sort of pollution ending up in the waters of the planet, many people think of the Mississippi River and the Gulf of Mexico. This is understandable, because so much farming pollution from the Midwest ends up in the Mississippi River, compounding the environmental damage done by the petroleum and natural gas industries along the shores of and in the Gulf of Mexico.

What goes around

This sort of water pollution and marine life death isn't only happening in the Gulf of Mexico, it is happening in the rivers and lakes and along and off the coasts of every continent, and among island nations where the use of farming, gardening, and landscape chemicals are increasingly being used.

Feed crop chemical pollution

When fertilizers from farms, golf courses, campuses, sporting fields, lawns, and other landscapes enter the ocean, they spur algae growth that blankets and kills the coral reefs. In the natural balance, fish and other marine life would consume the algae, but with the double threat of overfishing and pollution, the algae has gotten out of control. Because marine life populations have plummeted, there aren't enough fish or oysters and other filtering creatures to consume the amount of algae that would naturally occur without fertilizers. The fertilizers are causing so much algae growth that even an abundant and healthy population of marine life couldn't keep up with the growth.

Remaining coral reefs are growing at stunted rates, or they are dying.

More worrying picture about the plight of the oceans

"What we have seen here is something different. When you look at all of the individual impacts, and then you look at how they interact, you see a bigger story and a more worrying picture about the plight of the oceans. What we are seeing is that the individual impacts interact, and that the magnitude and scale of problems for the oceans is so much greater than what we probably originally thought. What we are also seeing is some of the changes that were perhaps predicted for the future are happening much, much sooner than we originally thought. And it is that combination that I think is really a wake-up call for all of us in terms of how we need to better manage the ocean, have better governance for the ocean, and make decisions with more ambition, and more quickly."

– Dan Laffoley, Senior Advisor at the International Union for the Conservation of Nature; June 2011

Life around reefs decreasing

Coral reefs are supporting only a fraction of the marine life they hosted just a few decades ago. In addition to overfishing, development of coastal cities and oceanside resorts, boating, military activity, oil drilling, fertilizers, industrial pollution, and the related algae and bacteria overgrowth, the death of coral reefs is caused by the amount of carbon dioxide being absorbed by the oceans.

Carbonic acid and exoskeletons

In ocean water, carbon dioxide forms carbonic acid. An abundance of carbonic acid decreases carbonic ions. The carbonic ions are key components of the exoskeletons of the tiny polyps forming the coral reefs. Under natural circumstances, the polyps cling onto the reef while feeding from the water. As the polyps die, their skeletons become part of the reef, and other polyps cling to the layers of skeletons. But when their systems aren't able to absorb enough carbonic ions, the polyps can't survive, don't reproduce, and the coral reefs don't grow.

Algae overgrowth among reefs

When algae overgrowth sets into a coral reef, and when the marine life is not at healthful levels, the coral reefs are subjected to bacteria overgrowth. When the reefs are in balance, the marine life consumes the bacteria and algae as natural sources of food. But the marine life among most of the world's coral reefs is no longer in balance. This provides the terrain for bacteria overgrowth among the coral reefs. The bacteria feast on the sugars released by the algae. The bacteria also feast on the eggs and sperm of the coral polyps, and also consume the polyps. This also contributes to the loss of coral reefs.

When coral reefs die off, so too does all marine life dependent on them.

Coastal erosion

The death and collapse of coral reefs leaves coasts susceptible to erosion. When the coasts are damaged, wildlife dependent on the coastlines lose their homes. Many coastal areas, including kelp forests, mangrove forests, seagrass lagoons, tide pools, delta wetlands, salt marshes, and the mouths of rivers are the breeding areas and nurseries for fish and other waterlife.

The acidification of the oceans is only part of the problem faced by coastal waterlife.

Coastal Nature being destroyed

Throughout coastal areas of the world the mangrove swampland forests have been and continue to be destroyed to make way for resorts, piers, bridges, roads, parking lots, marinas, gas refineries, shipping ports, shopping villages, entertainment and sports venues, power plants, and for shrimp farms, and other types of seafood farms. As the coasts are destroyed, so too are kelp and seagrass beds, wetlands and swamps, and the forms of life dependent on them.

The destruction of the coasts contributed greatly to the damage done by the tsunami that swept hundreds of thousands of people from villages and towns throughout the Indian Ocean region in December 2004.

Deepwater Horizon

"I flew over the Deepwater Horizon site with the Coast Guard. As we cleared Barataria Bay and Grand Isle, the first thing I noticed was how completely the oil industry has taken over the ocean. I could see dozens upon dozens of structures: some of them just small squares, others tree-like risers, a few big enough to accommodate helipads. All part of the 4,000 oil and gas platforms that make up the Gulf's industrial archipelago.

Equally shocking was the tattered coastline. From the air, the erosion of the marshes was impossible to miss. Lines of dead tress poked up from the sea and one-time channels were now surrounded by open water, berms on each side marking where the grasses once had been. The landscape looked like a watery skeleton."

– Riki Ott, *Hide and Leak: BP's Cleanup Is More Like a Cover up. Holding the Company Accountable Will Require Digging for the Truth; Earth Island Journal,* EarthIslandJournal.org; Autumn 2010

Clearing of coastal forests

Clearing of coastal forests and damage to marshes and barrier islands to create fish, oyster, shrimp, and other seafood farms, and to drill for oil and gas, also contributed to the strength of Hurricane Katrina, which decimated the New Orleans region in 2005.

Ongoing crude oil problem in Gulf

The mess that became the oil gusher beneath the Gulf of Mexico during 2010 remains a problem as much of the raw crude and the chemicals dumped into the Gulf to disperse the crude continues to kill a wide variety of sea life. Even all of these years later.

Fukushima and radioactive isotopes

As mentioned earlier, the Fukushima nuclear disaster that occurred in Japan in 2011 continues to play out. Large quantities of water polluted with radioactive isotopes continue to end up in the soil, ground water, and the ocean. It's something new humans have nothing much to compare to. It might lead to a variety of problems never experienced among such a wide variety of people and wildlife.

The enemy is us

"Although the Associated Press news editor's poll declared the BP disaster the biggest news story of 2010, much of the media still doesn't underline the fact that we burn over half of America's 20 million-barrel-a-day oil consumption in automobile fuel tanks. The media did an admirable job covering the drama of the BP blowout, and the debacle of the 'cleanup,' but it completely failed to explore the root causes of the problem. Until the medial begins connecting the dots, we, the people – like addicts everywhere – keep hoping that somehow technology or events, or something besides our own responsibility will save us. As the cartoon Pogo once put it: 'We have met the enemy and the enemy is us.'"

– Randy Salzman, *Still Crazy After All These Years; Earth Island Journal*, Spring 2011

Nearly two centuries of burning fossil fuels

The damage to marine life we have become aware of in the recent decades is likely an example of what we will see in coming years. This is because the oceans keep absorbing the carbon dioxide and other pollution that has been hanging in the atmosphere for decades, and which is the result of the burning of petroleum, coal, and natural gas. As the carbon dioxide from burning fossil fuels over the past couple centuries keeps being absorbed by the water of Earth, the lakes and oceans will increase in acidity, and temperature. Wildlife will suffer, and populations of both water and land animals will continue to decrease.

What humans have done to Earth, wildlife, plants, and the circle of life in two hundred years is astounding, and tremendously awful.

Air pollution and human disease

Air pollution increases the incidence of lung diseases, liver disorders, and heart disease, strokes, and a variety of degenerative and chronic health problems in people and wildlife. Chief culprits in air pollution are the extraction and burning of fossil fuels, deforestation and desertification, and the animal farming industry and its products.

It is only recently that the increase in ocean acidity is becoming widely known.

Coral shells dissolve

Scientists have discovered that coral and shells dissolve when ocean water becomes too acidic. This specifically presents a potentially devastating situation for all forms of shell-based waterlife, for the structures naturally left over as the shells gather, and for all forms of sea life dependent on those shelled creatures.

The acidification of the oceans is not only killing off and collapsing coral reefs, but also is destroying small islands (atolls) comprised of ancient coral reefs and tidal collections of old shells. Many small island nations consist of atolls. These are at increased risk of collapse.

Island nations

Many island nations depend on shellfish and healthy coral reefs to supply their food. Without healthy coral reefs supporting a number of small creatures that support the bigger fish, the people of these islands are at risk of losing their major food source. Many of them do not have the land to grow food, and lack the finances to import food from other islands or from the continents. Nor do many have the type of financial or coastal situation allowing for fish farming (a practice problematic for the environment and wildlife, and requires tremendous amounts of food to feed hundreds of thousands to millions of confined creatures.)

Food insecurities growing

"In the next 10 to 50 years many countries are going to see impacts, particularly countries heavily reliant on clams and oysters and mussels, and will not be able to adapt by shifting to other foods or aquaculture [fish farming] methods.

The truth is, if you look at all the effects of ocean acidification, nobody really gets off the hook. Impacts on coral reefs, for example, affect tourism, and affect fisheries – because fish depend on coral reefs. And when you look at those impacts, what you find is, in many cases, it's the developed countries, like the United States, the U.K. and other European countries that will be impacted.

If all these countries are going to have food insecurities because their clams or oysters are no longer available, or because their fisheries are no longer available as a result of climate change, that could put pressure on other countries – and it can affect all of us."

– Jackie Savitz, senior scientist and chief strategist for the international ocean conservation and advocacy organization Oceana, na.oceana.org; July 2011

Island collapse

As the oceans become more acidic by continuing to absorb carbon dioxide and industrial pollution from past and future decades, to where will the island people move when their homes flood and the islands collapse? Will the people still be considered a country when they lose their land? What will happen to their family dynamics, cultures, and languages?

"Anything that secretes limestone and it's shells, various forms of plankton, sea urchins, starfish, a lot of things – not only corals – extract the limestone from the sea, to deposit, to make their exoskeletons, and that requires more energy, and they can't do it beyond a certain point. And if you add to that, other impacts, it just adds more stress to them.

If you want to use a terrestrial, or human analogy, it might be that you can easily survive a disease, like influenza, measles, etcetera like that, one after the other. But, if you have them all at once, people usually don't survive. People can't cope with a lot of impacts all at once. And the impacts we are talking about with reefs are warming, sewage, overfishing, acidification, and extraction of resources in a way that push the reef beyond the lethal threshold at which they can survive."
– Charles Sheppard, Warwick University; June 2011

All shellfish play into the circle of life

Coral reefs aren't the only forms of sea life dependent on an abundant supply of carbonic ions. All shellfish need carbonic ions to build their calcium carbonate skeletal structures. These include clams, crabs, lobsters, shrimp, krill, urchins, barnacles, and oysters. These crustaceans and mollusks provide food for otters, seals, walruses, seabirds, and other types of wildlife. In turn, those creatures support the life of other creatures. Even the lichen, grasses, ferns, mosses, shrubs, and trees along the coasts are dependent on the health of the smaller forms of life in the oceans, this is because when the seabirds, and land mammals living around the coasts leave their droppings, and when their bodies decompose, the nutrients in the droppings and the decomposing bodies end up as nutrients for fungi, bacteria, bugs, raptors, and other forms of life living on or near the coasts – which then, in their network of life, spread the nutrients even wider, on the land. All of those nutrients help grow our food, forests, and other things we depend on to survive.

Ecocide on the largest scale

Humans are dependent on a rich variety of sea life, including shelled creatures. The more fossil fuels humans use, the more humans are ruining their chances of survival. It's ecocide on the largest scale.

Predatory fish biomass

"We estimate that large predatory fish biomass today is only about 10% of pre-industrial levels.
We conclude that declines of large predators in coastal regions have extended throughout the global ocean, with potentially serious consequences for ecosystems."
– Ransom Myers and Boris Worm in *Nature*, 2003

We've changed Earth's temperature

"We've actually changed the temperature in the last century just the right amount. The West Antarctic ice sheet is starting to melt.

We can expect a sea level rise of one to two meters by the end of this century, and it could be larger than that. This is a serious consequence for nations like Kiribati, where the average elevation is about a little over a meter above sea level."

– Rob Dunbar, oceanographer, biologist

Arctic cod

The Arctic cod – which feeds on zooplankton and smaller fish living beneath the sea ice – have been suffering as the sea ice has been disappearing. With less sea ice, and more fishing fleets have access to the formerly ice-covered waters. Because of this, the Arctic cod are even more susceptible to overfishing.

Overfishing is one of many problems facing ocean life.

As the Arctic cod have to search more for their food, their bodies use up more energy, and the fish are not the size they used to be. The cod naturally stay in the colder waters, heading north during the summer months, and south during the colder months. With the ice melting and the waters remaining warmer throughout the year, the cod are staying more north. With fewer varieties of food for the cod to consume, the cod populations have been decreasing.

"In the Arctic cod's favor: A fishing ban. To prevent a free-for-all in newly ice-free waters, the North Pacific Fishery Management Council in 2009 banned all large-scale commercial fishing in American territorial waters above the Bering Strait. Perhaps most important, the ban covers not just fish but 'all other forms of marine animals and other plant life.' That may forestall the kind of trouble now brewing in the waters around Antarctica, where a fast-growing industrial krill fishery [it's sold as food for farmed salmon and pressed into oil for omega-3 supplements] threatens the base of the food chain in the Southern Ocean [Antarctic Ocean]. Krill are less abundant in the Arctic, but the growing demand for the crustaceans could lead krill processors to turn their eyes to northern waters."

– Bruce Barcott, *Arctic Fever*, OnEarth, the quarterly magazine of the Natural Resources Defense Council; NRDC.org; Spring 2011

Marine microlife

While people know about the migration of birds, butterflies, whales, and other wildlife, most don't know about the largest migration of wildlife on the plant, which takes place every day. This migration involves the tiny beings living in ponds, lakes, and oceans. It is a vertical migration in which trillions and trillions of amphipods, copepods, diatoms, foraminifera, mini jellyfish, and other zooplankton take part. As the sun sets, the zooplankton rise from the depths, and as the sun rises, the zooplankton descend, away from the ultraviolet radiation. As they travel, some become food for larger forms of life, including fish, sea mammals, insects, amphibians, and birds, and in that way they support all forms of waterlife. Probably the most well-known of the smaller forms of waterlife are krill.

As mentioned earlier, krill are taken from the water and fed to farmed fish. The health food markets ignorantly sell "krill oil" as a nutrition supplement. As demand for krill has increased to satisfy demand for the shrimplike creatures to feed to the growing farmed fish industry, and to fulfill orders for the krill oil supplement businesses, the krill populations have greatly decreased. This threatens the creatures that feed on krill, including whales, seals, penguin, and other marine birds, and fish – which, in one form or another, are food for other creatures.

Plankton

The microscopic plankton, which are the food for whales and krill, also have calcium carbonate shells. If the plankton populations decrease, the whales and krill-eating seabirds will die. Plankton levels in the oceans today are at an all-time low. This is directly related to global warming and the acidification of the oceans, which is directly related to the use of fossil fuels, including the use of natural gas to create synthetic fertilizers which, like other industrial chemicals being used around the planet, end up in the swamps, rivers, lakes, marshes, and oceans.

Oceanic perspective

"Now, you have heard about two degrees, that we should limit temperature rise to no more than two degrees. But, there is not a lot of science behind that number. We've also talked about concentrations of carbon dioxide in the atmosphere, should it be 450? Should it be 400? There's not a lot of science behind that one, either. Most of the science behind these numbers, these potential targets, is based on studies on land. And, I would say, for the people that work in the oceans, and think about what the targets should be, we would argue that they must be much lower. From an oceanic perspective, 450 is way too high. Now, there is compelling evidence that it really needs to be 350. We are, right now (2010), at 390 parts per million of CO_2 in the atmosphere. We're not going to stop in time to put the brakes on at 450, so we've got to accept that we are going to do an overshoot, and

the discussion as we go forward is to focus on how far the overshoot goes, and what is the pathway back to 350."
– Rob Dunbar, oceanographer, biologist

Whales facing difficulties

Whales are facing increasing environmental and food source difficulties. Many whales have died after colliding with ships or tangling in fishing nets and ropes. More have died because of hunting. Others are dying from plastic and other forms of pollution. Their body chemistry is being altered by pollution, including pharmaceutical pollution. And their food sources are decreasing.

Tradition of high seas freedom

"The tradition of freedom of the high seas has its roots in an era when there were too few people to seriously violate the oceans – but in hindsight that era ended some 150 years ago with the increase of global whaling."
– Jim Carlton and Claudia Mills in speech to University of Washington School of Marine Affairs, 1998

Hunting whales

Against international law, Japan, Norway, and Iceland continue to allow their citizens to hunt whales. And they continue to ignore environmental and animal rights activists, and scientists.

The Japanese whaling community hunts piked, mink, sei, Bryde's, and fin whales. According to SeaShephard.org, the Japanese hunters also remain a threat to humpback whales.

Japan's whaling industry is partially financed by the government, as is the hunting of dolphin.

"Japan is dead set on hunting whales – international critics be damned. In 2014, the International Court of Justice ordered the country to stop hunting whales. But at the beginning of this month, Japan announced it would send a small whaling fleet into the Antarctic Ocean to kill 333 whales under the guise of a scientific program.

Unlike Americans, Japanese people don't tend to see whales as charismatic mammals that should be protected from human consumption by a universal taboo."
– Tom Philpott, *2 Surprising Reasons Why Japan Won't Stop Hunting Whales*; *Mother Jones*; Dec. 2015

"We do not accept in any way, shape, or form the concept of killing whales for so-called scientific research."
– Greg Hunt, Australian Environment Minister

"Japan makes no secret of the fact that the meat resulting from its so-called scientific whaling program ends up on the plate."
– BBC

"On July 1st 2019, Japan resumed commercial whaling after leaving the International Whaling Commission (IWC).

In 2021, Japanese whaling vessels hunted a self-allocated quota of 171 mink whales, 187 Bryde's whales, and 25 sei whales.

Japanese whalers also undertake hunts of dolphins, such as those at Taiji, which are not subject to IWC rules.

The Japanese government has clearly and repeatedly stated that they want to return to commercial whaling, despite the opposition of most of the world's countries, and a shrinking market."
– *Whaling in Japan*; US.Whales.org

Sea Shepherd's fight against whaling

Due to the efforts of Sea Shepherd, in February 2011, the Japanese government whaling ships gave up on their Antarctic whaling for the season. This was a victory for those who worked to stop the ships from killing more whales. Rather than reaching their goal of killing 850 minke and 50 fin whales, the fleets had killed 170 minke and 2 fin whales – still a tragedy, but far below what it could have been.

Thanks in part to the international attention given to the cruelty, horror, and destruction of whaling, many Japanese people have stopped eating whale meat. Eating whale meat hadn't been common in Japan until after World War II. Claiming it is a tradition thousands of years old is in error. Most Japanese people don't eat whale meat.

Whales on the edge of extinction

"Many whales are and were on the edge of extinction. As a result, the International Whaling Commission established a moratorium against commercial whaling in 1986, except for research activities. The so-called 'research loophole,' has made the whales pay a heavy price. The Japanese have killed thousands of whales since the moratorium under this pretense, stockpiling ten million pounds of frozen meat for sale, and introducing it to school cafeterias. But some goes for pet food. It has been proven non-lethal means of research are scientifically more effective, but the Japanese are not interested. They want the meat.

An incredibly brutal method of killing has been directed at the most intelligent and social mammals on Earth: shot with exploding harpoons that rupture organs in their bodies, they are then cruelly winched in where probes are jammed into their flesh, electrocuting them with thousands of volts for 15 minutes on end, until they finally succumb – or are dragged through the waters by their tales until they drown. Such a practice would never be condoned anywhere on land."

– From the documentary, *The Whale Warrior: Pirate for the Sea*; pirateforthesea.com

Whales and other sea animals face incredible odds of survival.
Then, there are land creatures bred and raised to be killed and eaten.

Chicken hell

"Chickens are probably the most highly consumed protein source on Earth, and also the most badly abused farm animals with the worse management practices. But we still turn a blind eye to their treatment, factory farming abuse, and the cruel slaughter process.

According to statistics of 2023, there are almost 34.4 billion chickens in the world.

The U.S. is the leading country of chicken production, with 9.22 billion chickens."

– World Animal Foundation, 2023

Billions upon billions of farmed animals

While the wildlife of the rivers, lakes, seas, and shores continue to be treated as food sources to be caught and farmed, killed, and eaten, the non-farm wild animals, especially predator animals, continue to be in danger of extinction. This is mainly because of cattle ranching, monocropping, road building, fracking, mining, and city and town sprawl. While this is happening, farm animals, including cows, hogs, chickens, turkeys, goats, and lambs, are being bred by the billions to feed the increasing demand for meat. To support those billions of farmed animals, a large majority of the land being farmed on the planet is used to grow food for farmed animals.

Damage of meat diet

"Americans' appetite for meat and dairy – billions of pounds every year from billions of animals – takes a toll on our health, the environment, the climate, and animal welfare. Meat and dairy production requires large amounts of pesticides, chemical fertilizer, fuel, feed, and water; and generates greenhouse gases, toxic manure, and other pollutants that contaminate our air and water."

– Environmental Working Group; July 2011

Animal farming presents a variety of environmental disasters. Among those are the damage animal farming and its network of support industries cause to ponds, swamps, streams, rivers, lakes, marshes, and oceans. As mentioned, growing massive quantities of food to feed billions of farmed animals on every continent and many islands decimates micro ecosystems, causes the extinction of species, changes the climate, drains aquifers and lakes, redirects or eliminates rivers, saturates the soil with farming chemicals, which end up in the rivers, lakes, and oceans, and eliminates wildlife from their natural habitat that is turned into vast industrial monoculture farming operations.

Concentrated animal feed operations: factory farms

"In Iowa, home to hundreds of CAFOs [concentrated animal feeding operations = factory farms], the Department of Natural Resources recorded that 99 waterways were contaminated enough in 2008 to cause fish kills, and that 47 of the incidents that caused the contamination could be positively traced back to animal waste. Such contamination has killed as many as 150,000 aquatic animals at a time."
– Monica Eng, *The costs of cheap meat. Critics of factory farms say we pay a high price for low-cost food*; Chicago Tribune; Sept., 2010

Monocropping feed crops

When thousands of acres of land are cleared and altered to monocrop food for the world's billions of farmed animals, that land is also cleared of the animal life that existed there for thousands of years; destroys stream beds and the life dependent on those streams; decimates native plants; reduces the variety of life forms in the region; damages the soil; reduces the chances for the soil and the organisms that had been living in it to regenerate; and even changes the atmospheric conditions in the surrounding region – especially when forests are cleared, grasslands are reduced, wetlands are drained, and streams and rivers are damaged, dammed, destroyed, or otherwise impacted.

"The problem is there is very little monitoring of the pollution big ag has caused because the agencies don't regulate."
– Scott Edwards, Waterkeeper Alliance.

Chesapeake Bay polluted by animal farming

"An analysis by the Chesapeake Bay Program found that agriculture – both livestock and crops – is the single biggest source of pollution in the bay, contributing 42 percent of the nitrogen, 46 percent of the phosphorus and 76 percent of the sediment in the troubled waterway."
– Monica Eng, *The costs of cheap meat. Critics of factory farms say we pay a high price for low-cost food*; Chicago Tribune; Sept., 2010

What is the chief way you can avoid supporting the clearing of land of native plants and animals to grow food for farmed animals? Become vegetarian. More specifically, follow a vegan diet. The more organic, plant-based, and local your diet, the better.

Human disease and meat and dairy

"Americans consume more meat and dairy than any other country in the world, and have the high rates of colon cancer, heart disease, and diabetes to prove it."
– Dr. Andrew Weil

The human dietary requirements for animal protein, including meat, milk, and eggs are absolutely zero.

While many people live in regions of the planet where growing food 12 months out of the year isn't possible, the majority of people in industrial countries, which are the people who cause most of the pollution on the planet, do have the option of getting a wide variety of fruits, vegetables, nuts, seeds, and seaweeds throughout the year.

Most of the resources

Most of the resources being used on the planet are to support the meat and dairy industries and to prepare and package meat, eggs, and dairy for shipping, sale, cooking, and consumption. This includes all of the materials, land, water, and fuels used to support the fishing and animal farming industries – including steel, packaging, shipping vehicles, stores, warehouses, stoves, ovens, and refrigerators.

"It is estimated that in the past thirty years, the human species alone has used up one-third of the entire planet's natural resources... over-fishing, in all parts of the world, is just one example."
– ZeroImpactProductions.Com

Going veg

The human body can flourish in health by consuming a vegan diet rich in a variety of unprocessed plant matter, and free of chemicals, junk food, fried foods, extracted oils, and clarified sugars. By utilizing sites like ForksOverKnives.com and NutritionFacts.org, and reading books such as those by Dr. Michael Klaper, Dr. Caldwell Esselstyn, Dr. John McDougal, Dr. Neal Barnard, and Dr. Dean Ornish, and my book *Plant-based Regenerative Nutrition*, a person can learn ways of following a diet free of meat, dairy, and eggs. This will greatly benefit human, environmental, and wildlife health.

Arthritis and vegetarian diet

"For many years, researchers have been studying how foods affect arthritis. Some of the early studies were not of the best quality. But by 1991, the issue was settled beyond any reasonable doubt. In *The Lancet*, a prominent British medical journal, researchers reported that a specially designed vegetarian diet can greatly reduce the signs and symptoms of arthritis. In the study, researchers found that a vegetarian diet lessened joint stiffness, swelling, and tenderness, and improved grip strength. The benefits lasted long after the study was over.

Here is why it works: Certain foods act as an arthritis trigger, stimulating the inflammatory process that attacks the tender synovial lining that is inside joints. The most common trigger foods are dairy products. Because it appears that a dairy protein is the culprit, even fat-free versions can trigger the inflammation that causes pain. A switch to soymilk or rice milk can help. Eggs and meat can contribute to joint pain for some people, which is why researchers are especially fond of vegan diets."
– Dr. Neal Barnard, author of *Foods that Fight Pain*

Don't eat anything with a face

"By eating these whole foods, and getting away from processed foods, getting away from the dairy, and anything with a mother, anything with a face – beef, pork, other mammal meat, and fish, and birds – it's incredible how powerful the body can be. If we are going to have a seismic revolution of health in this country, which is really right at our fingertips, then the major behavior that has to change is our food [intake]. That is absolutely the key card, it trumps everything."
– Dr. Caldwell B. Esselstyn, author of *Prevent and Reverse Heart Disease*

Seafood

Many people think vegetarians eat fish. Some people who consider themselves to be vegetarians do eat fish, and some who say they are vegetarian also eat birds. Those people aren't vegetarian.

A vegetarian does not eat fish, birds, mammals, reptiles, amphibians, or any part of any animal of any sort.

Seafood is bush meat

"Seafood is simply a socially acceptable form of bush meat. We condemn Africans for hunting monkeys and mammalian and bird species from the jungle, yet the developed world thinks nothing of hauling in magnificent wild creatures like swordfish, tuna, halibut, shark and

salmon for our meals. The fact is that the global slaughter of marine wildlife is simply the largest massacre of wildlife on the planet."
– Captain Paul Watson, SeaShepherd.org

Fishing damage

While many people are aware of the damage to water life caused by pollution, most people are not aware of the damage to the oceans, lakes, rivers, and streams caused by fishing. It is because of fishing that many types of sea life have disappeared, and others are being seriously threatened with extinction. Many populations of sea life the fishing industry seeks are at dangerously low levels.

The term "seafood"

• The term "seafood" includes hundreds of species of water-dwelling creatures, from fresh and saltwater fish, to crustaceans (lobster, prawns, shrimp, etc.), and mollusks (clams, squid, etc.) to cetaceans (water mammals: whales, dolphins, porpoises, seals, etc.).

Per-person consumption of seafood

• The average international per-person consumption of seafood has nearly doubled in the last half of the 20th century. Part of that is because the popularity of refrigeration, and the ability to transport seafood by airplane.

Americans eating 19 pounds of fish per person per year

• According to the National Oceanic and Atmospheric Administration of the U.S. Department of Commerce, in 2004 the average American ate 16.6 pounds of seafood. The figure indicated a 1.4-pound increase in four years.
  According to The National Oceanic and Atmospheric Administration, in 2020 Americans were eating about 19 pounds of fish per year. (Not me).
  In 2022, Americans were still eating about 57 pounds of beef and about 68 pounds of chicken per person. Many people eat far more. Many people consume some form of meat at every meal.

Land animal consumption

• In 2005 the consumption of land animals (cows, pigs, chickens, turkeys, ducks, lamb, goats, etc.) was 2.5 times greater than of seafood.

Shrimp consumption

• In 2022 the average American consumed 5.9 pounds of shrimp.

Tuna consumption

- In 2022 there were over 1.2 billion cans of tuna sold in the U.S.

Fish farms

- Seafood taken from the world's waters, including fish farms, has gone from about 20 million tons in 1950 to about 130 million tons in 2000. By 2023 it had risen to 200 million tons.

Sea life fed to farmed animals and used as fertilizer

- About a fourth of the sea creatures taken from the oceans aren't used for human consumption, but are instead turned into pet food, into feed for farm animals (mostly fed to pigs, chickens, and turkeys), into feed pellets for fish farms, and into fertilizers. This includes everything from krill to shrimp and fish not wanted by the markets, and other forms of sea life.

China's seafood consumption

- China, with only four percent of the world's coastlines, is the world's lead consumer of fish, and produces over 28 million tons of farmed fish yearly. China consumes about 38% of the world's global consumption of fish.

Refrigeration increased seafood consumption

- Advances in refrigeration combined with the building of airports around the world have made it possible for people in virtually any region to eat sea creatures from any other region. Before refrigeration, most people didn't eat much seafood.

Fishing is of the most dangerous professions

- Commercial fishing continues to be one of the most dangerous professions.
  According to the U.S. Centers for Disease Control and Prevention, more than 878 commercial fishermen died between 2000 and 2019, averaging over 43 deaths per year. Half were because of vessel disaster, and about 30% from falls overboard.

"Researchers say the global fish harvest will have to increase by nearly 50 percent by 2020 just to meet new demand in China and other developing countries."
  – Craig Cox, *The Great Open Ocean Sell-off: A Glut of New Offshore Factory Fish Farms May Be Just Over the Horizon*; *Utne* magazine; Utne.com; Nov.-Dec. 2004

Western Flyer and John Steinbeck

In 1940 a sardine fishing boat named The Western Flyer left the Monterey, waters, traveled down past the Baja California coast, around Cabo San Lucas, and up to the top of the Sea of Cortez. Among those on board were *The Grapes of Wrath* author John Steinbeck, and his marine biologist friend, Edward F. Ricketts, who wrote the landmark book *Between Pacific Tides*, a study of intertidal marine biology. The 4,000-mile six-week trip was recorded in Steinbeck's nonfiction 1951 book, *The Log from the Sea of Cortez*. The book stands as a survey of the sea life Steinbeck observed and encountered on the trip.

The week I am writing this, The Western Flyer is going to return to Monterey, to display the complete restoration it had gone through. As it hadn't been cared for in many years, in 2012 it had sank in the Swimonish Channel. Pulled from the water, it was transferred to a dock, where it spent years being restored.

The aquarium of the world

In his book, Steinbeck wrote about a sea filled with life, where sea turtles were common; where massive schools of groupers appeared, some weighing 500 pounds; where giant manta rays leaped from the water; where teams of yellowfish chased schools of sardines; where hammerhead sharks fed; where tuna jumped from the sea; where leaping swordfish cut through the water; where dolphin swam alongside the boat; and where pelicans dove to pick fish from the aquamarine waters. His descriptions of the coast of Baja portray a quiet and unspoiled land with random villages of Native people. The coastal waters had plentiful amounts of a variety of vertebrates and invertebrates. Pearl oysters were common, as were starfish, sea cucumbers, jellyfish, crabs, stingrays, snails, barnacles, anemones, and urchins.

Even as late as 1986 internationally renowned oceanographer Jacques Cousteau declared the Sea of Cortez to be the "aquarium of the world."

Web of ocean life crashing

"The web of life in our oceans is being pulled apart as top predators like tuna have crashed from overfishing; trawling for groundfish and scallops has scraped some areas of the ocean bottom clean like a parking lot; and pollution and nutrient-rich runoff have fed algae blooms and jellyfish population explosions, resulting in what one scientist calls 'the sliming of the oceans.'"
– Jon Christensen, Steinbeck Fellow, San Jose State University, *Back to the Sea of*

Cortez: Sailing with the Spirits of John Steinbeck and Ed Ricketts on a New Journey of Discovery Around Baja California, SeaOfCortez.Org

Retracing Steinbeck's Sea of Cortez travels

In 2004, Chuck Baxter, a retired Stanford University marine biology professor; Jon Christensen, a freelance writer (and keeper of the blog SeaOfCortez.Org); William F. Gilly, a neurobiologist; and Nancy Packard Burnett, who, with Baxter, is a cofounder of the Monterey Bay Aquarium, gathered a group of people together and retraced the 1940 voyage taken by Steinbeck and Ricketts. They used the writings of Steinbeck and Ricketts as guides while making the trip on a shrimp trawler named Gus D.

Sea of Cortez life has plunged

What Baxter's group found was vastly different from what Steinbeck recorded in his book, and what Ricketts wrote about in his 2004 field notes. Not only were many types of sea life described in 1940 by Steinbeck and Ricketts not seen, but even many of the creatures in the tide pools were missing, or in obvious steep decline.

Baxter, Christensen, and the others did not see turtles, sharks, schools of tuna, leaping swordfish, or giant manta rays. Over the decades, those and many other forms of life once common in the Sea of Cortez, such as tuna, shrimp, and yellowtail have vanished, or have been brought to the edge of extinction in those areas because of overfishing and pollution. In addition, much of the shellfish, including the pearl oysters, have been hunted to death. The sea life ended up in grocery stores and restaurants, as pet food, as feed for farmed animals, as fertilizers, and as tourist junk. As time passed much of the coast of Baja has been built into tourist traps, retirement communities, college student party zones, and harbors for yachts and cruise ships for the visiting rich.

Fishing industry "large, destructive machine"

Steinbeck had written about the fishing industry, which he called "a large, destructive machine," and which was involved in "committing true crimes against Nature." What the group traveling in 2004 saw was the result of those crimes: a sea with relatively few forms of life left, compared to just 64 years in the past.

What has happened to the Sea of Cortez has happened and is happening to oceans and seas, lakes and streams all over Earth. Pollution from urban runoff, from the oil industry, from the burning of fossil fuels, from the shipping industry, and from such things as cruise ships, the military, golf courses, and the chemicals used in the farming industry has resulted in poisoned waters and "carbon sink-holes" where oxygen is so depleted in the dead zone areas that natural sea life cannot exist where it had been for many thousands of years.

Sea of Cortez to Chesapeake Bay: loss of marine life

"The Chesapeake Bay, on the East Coast of the United States, has one of the first dead zones ever identified, in the 1970s. The Chesapeake's high levels of nitrogen are caused by two factors: urbanization and agriculture. The western part of the bay is full of factories and urban centers that emit nitrogen into the air. Atmospheric nitrogen accounts for about a third of the nitrogen that enters the bay. The eastern part of the bay is a center of poultry farming, which produces large amounts of manure."
– Dead Zone, *National Geographic Encyclopedia*, 2023

Chesapeake Bay, which, like the Sea of Cortez, was once compared to a giant aquarium, is also void of many forms of sea life that once flourished there. Instead, Chesapeake Bay is highly polluted from the runoff of cities, animal farms, and chemicals used on farms, lawns, corporate and government landscapes, sporting fields, golf courses, and along roadways.

Strangling the bay

The marine life of Chesapeake Bay is being strangled by water-clogging algae feeding off the nitrogen in the fertilizers and farm runoff. The bay is the world's main spawning ground for striped bass, which are often found to be starving and with bacterial infections eating away at their flesh. The natural water-filtering sea creatures, including oysters and menhaden fish, are on the steep decline caused by pollution and fishing, and their absence is adding to the pollution.

Dead zones in Gulf of Mexico

There are vast areas of dead water covering several thousand miles of water in the Gulf of Mexico, where the natural sea life has died off and is unable to exist. Throughout the world's seas there are more and more dead zones lacking oxygen, and poisoned by pollution.

Water Hypoxia Task Force

"Low river discharge is a key factor contributing to this year's dead zone.

Today, NOAA-supported scientists announced that this year's Gulf of Mexico dead zone – an area of low to no oxygen that can kill fish and marine life – is approximately 3,275 square miles. That's more than 2 million acres of habitat potentially unavailable to fish and bottom species – larger than the land area of Rhode Island and Delaware combined.

The five-year average dead zone size (also known as the hypoxic zone) is now 4,280 square miles, which is over two times larger than management targets. Since records began in 1985, the largest hypoxic zone measured 8,776 square miles in 2017.

The measurement was made during an annual survey cruise, led by a team of scientists from Louisiana State University and the Louisiana Universities Marine Consortium (LUMCON) during the final week of July. The information gathered is a key metric used by the Mississippi River/Gulf of Mexico Water Hypoxia Task Force to measure progress toward achieving their five-year average target of 1,900 square miles or smaller by 2035.

In June, NOAA forecasted an average-sized hypoxic zone of 5,364 square miles, based primarily on Mississippi River discharge and nutrient runoff data from the U.S. Geological Survey.

Each year, excess nutrients from cities, farms, and other sources in upland watersheds drain into the Gulf of Mexico and stimulate algal growth during the spring and summer. The algae eventually die, sink, and decompose. Throughout this process, oxygen-consuming bacteria devour the algae and consume the oxygen. The resulting low oxygen levels near the bottom are insufficient to support most marine life, rending the habitat unusable, and forcing species to move to other areas to survive.

Exposure to hypoxic waters has been found to alter fish diets, growth rates, reproduction, habitat use, and availability of commercially harvested species like shrimp."
– *Below-average Gulf of Mexico 'dead zone' measured*; National Oceanic and Atmospheric Administration, NOAA.com, Aug. 2022

The goal is to keep the Gulf of Mexico dead zone to be no larger than 1,900 square miles. Although, ideally, the dead zone would not exist.

"While some hypoxia is natural, the size and scale of what we've seen here in the last several decades is unusually large and detrimental. Our measurements and analyses can empower communities to take action to protect their coasts and contribute to the region's economic stability."
– Nicole LeBoeuf, director of NOAA's National Ocean Service

It is sad that there is even a "management target" of keeping any dead zone of any sea to be so large, and that a water hypoxia task force even needs to exist.

Extreme animal farming

Because of extreme animal farming practices, excrement from the animals, and the use of chemicals on millions of acres of feed crops, and other pollution, dead zones exist in oceans and lakes around the planet.

In 2018, the largest water dead zone was believed to be in the Arabian Sea.

Arabian sea dead zone

"The Gulf of Mexico's dead zone is bigger than ever. Recent surveys put it at an enormous 8,776 square miles, large enough to cover New Jersey. But another massive zone of low dissolved oxygen confirmed recently in the Arabian Sea is seven times larger. At 63,000 square miles – the size of Florida – it ranks as the world's largest.

Scientists from the University of East Anglia of Norwich, England measured the dead zone, which sits in the Gulf of Oman south of Iran, with underwater robots. The area had been suspected of hosting a massive dead zone, but moving bands of pirates and the region's volatile geopolitics made research difficult. The torpedo-shaped robots were able to slip in and do the measurements with ease, but they came back with very bad news.

'The Arabian Sea is the largest and thickest dead zone in the world,' said Bastien Questo, a marine biochemist and the study's lead author. 'But until now, no one really knew how bad the situation was because of piracy and conflicts in the area have made it too dangerous to collect data.'

Waters depleted of oxygen turn fish away and suffocate anything that can't escape, including plants, slow-moving crabs, and other shellfish.

The Gulf of Mexico's dead zone is the largest in the Western Hemisphere. Before the size of the Arabian Sea's low oxygen was confirmed, Europe had what was considered the world's largest dead zone. Starting as a small patch of low oxygen off the coast of Sweden a century ago, the Baltic Sea dead zone has ballooned to more than 23,000 square miles, and now stretches from Poland to Finland. Nutrient pollution and rising deep-water temperatures are the main causes, according to the Stockholm University Baltic Sea Centre."

– Tristan Baurick, *World's largest dead zone discovered, and it's not in the Gulf of Mexico*; The Times-Picayune, May 2018

Landscape and farming chemicals and animal disease

Some of the things communities can do to reduce pollution flow into the network of rivers emptying into the Gulf of Mexico include banning the use of toxic chemicals on lawns, sports fields, government and corporate landscapes, and on feed crop fields. That would be wise, including to reduce rates of cancer, autoimmune disorders, lung disease, and other health issues among humans, their dogs and cats, and wildlife.

Organic farming and gardening

Another way to reduce pollution in the water is by NOT supporting the feed crop and animal farming industries, as the pollution from growing feed crops and from animal farming causes an enormous amount of pollution in ponds, rivers, lakes, swamps, marshlands, and oceans. In other words, follow a plant-based diet. If you have space, grow food organically. Also, support local organic produce farmers. Otherwise, eating meat and dairy only worsens the environmental damage caused by the animal farming and feed crop industries, including by causing – and increasing the size of – dead zones.

Drought and chemical buildup

Ironically, the drop in rainfall for a season can result in dead zones decreasing as plants and animals suffer from drought. This is so because fewer chemicals used on farms and landscaping then end up in water. But the buildup of the chemicals on the land, the filling of farm animal waste "lagoons," and building piles of farmed animal dung continues.

The next floods spread the pollution. The chemicals end up in the water, including the wells people drink from, the aquifers, and on into the rivers, lakes, and sea.

Factory farming waste lagoons

Some of the animal farming waste "lagoons" overflow, killing waterlife. It's mixed with farming chemicals. As it works its way into the water network, it helps to spur algae growth problematic for and kills marine life, including when the algae dies and bacteria feeds on the rot. Those are the dead zones. Or zones that are dead of what would naturally live there, as what is not natural overgrows.

Hundreds of dead zones, and counting

In 1995 the United Nations Environment Program identified at least 146 dead zones in the world's seas. By 2022, 415 dead zones had been identified. So far.

For the past seven decades, the number of global dead zones have been identified as doubling about every decade. With the ocean temperatures increasing, it increases the ability for mass quantities of algae to bloom, and that increases dead zones, killing sea life, including coral reefs, seagrass, kelp forests, and other areas where smaller fish and shelled creatures thrive.

Stressed marine ecosystems

"A seminal study by Professor Bob Diaz of the Virginia Institute of Marine Science (VMS) and Swedish researcher Rurger Rosenberg shows that the number of dead zones increased by a third between 1995 and 2007. Dead zones are now a key stressor of marine ecosystems and rank with over-fishing, habitat loss, and harmful algal blooms as global environmental problems.

Diaz and Rosenberg recorded 405 dead zones in coastal waters worldwide, affecting an area of 95,000 square miles, about the size of New Zealand. "
– David Malmquist, *Dead Zones: Lack of oxygen a key stressor on marine ecosystems*; Virginia Institute of Marine Science; Oct. 2021

Eutrophication, nutrient overdose, and cyanobacteria

"Dead zones occur because of a process called eutrophication, which happens when a body of water gets too many nutrients, such as phosphorus and nitrogen. At normal levels, these nutrients feed the growth of an organism called cyanobacteria, or blue-green algae. With too many nutrients, however, cyanobacteria grows out of control, which can be harmful. Human activities are the main cause of these excess nutrients being washed into the ocean. For this reason, dead zones are often located near inhabited coastlines.

Eutrophic events have increased because of the rapid rise in intensive agricultural practices, industrial activities, and population growth.

Human activities have emitted nearly twice as much nitrogen and there times as much phosphorus as natural emissions.

Different regions of the world emit different levels of these nutrients. In developed countries, such as the United States and nations in the European Union, heavy use of animal manure and commercial fertilizers in agriculture are the main contributors to eutrophication. Runoff from large agricultural fields enters creeks and bays because of rain or irrigation practices.

In developing countries of Latin America, Asia, and Africa, untreated wastewater from sewage and industry mainly contribute to eutrophication.

Atmospheric sources of nitrogen also contribute to eutrophication. Fossil fuels and fertilizers release nitrogen into the atmosphere. This atmospheric nitrogen is then redeposited on land and water through the water cycle (rain and snow)."
– Dead Zone, *National Geographic Encyclopedia*, 2023

Toxic soup zones and ocean acidification

Some dead zones exist near the surface of the water, and some are deep down. This, along with ocean acidification, increases the risk to a great number of marine species. It is one reason why sea life, such as sharks, and schools of fish, are found swimming where they didn't usually "normally" swim, and a main reason why some forms of sea fish and vegetation are vanishing from the oceans.

Nine of every ten organisms

"It is estimated that nine out of ten organisms on the entire planet live in the ocean. Some of these are tiny unicellular algae which make up part of the plankton that produce four times the amount of oxygen than terrestrial plants do."
– ZeroImpactProductions.Com

Humans: ending natural functions

The pollution of the rivers, lakes, and oceans is a result of the activities of one type of animal – the human.

Humans aren't only destroying the life and natural functions of the most distant parts of the oceans. They are destroying land and life forms in and on the most remote areas of the continents and islands. This is ecocide, and the end result could be omnicide. (Search those words.)

Intrusion of modern human society

From the deserts of the Middle East to the landscapes at the center of Australia to the ancient forests of South and Central America, humans are extracting fossil fuels to feed market demands for energy so people can drive, fly, cook, play video games, yap on cell phones, ride motorbikes on "nature trails," soak in heated pools, and consume what they absolutely do not need, meat, dairy, eggs, fried foods, and junk food.

The environment is being destroyed by drilling or/or mining for, and the use of fossil fuels.

Tar sands environmental destruction

"Just south of the polar region's barren tundra, Canada's boreal forest provides 1.3 billion acres of wild habitat for a mind-boggling array of species – from large carnivores like grizzly bears, wolves, and lynx to nesting migratory birds to thousands of plant varieties. Its trees and bogs capture vast amounts of climate-changing carbon pollution, and its wetlands filter millions of gallons of water. The boreal is also home to hundreds of First Nations communities, many of whom rely on hunting, fishing, and trapping for their livelihoods.

Yet this pristine paradise is central to one of today's most contentious debates about dirty energy. It's a conversation with origins dating back 300 years, when British explorer James Knight noted the 'gum or pitch that flows out of the banks,' of what is now Canada's Athabasca River.

Knight was describing tar sands, a sludgy deposit of sand, clay, water, and sticky, black bitumen (used to make synthetic oil) that lies beneath northern Alberta's boreal forests in a region the size of Florida. Extracting and converting tar sands into usable fuel is a hugely expensive energy- and water-intensive endeavor that involves strip mining giant swaths of land and creating loads of toxic waste, and air, and water pollution.

Despite these economic and environmental costs, a race to make money from this dirty fuel was kicked off in the mid-1990s by rising oil prices. By 2004, Canadian production of tar sands oil had reached one million barrels per day – with much of the output bound for the United States.

First Nations communities who had inhabited the new ground zero for tar sands oil development for thousands of years began voicing concerns. Their water, fishing, and hunting traditions were at risk – as was their health, with more rare and unusual cancers being diagnosed.

In 2005, these communities invited Canadian activist groups and NRDC to Alberta to talk about tar sands. They shared disturbing images of a lunar-like landscape of open-pit mines and vast wastewater containment ponds where boreal forest had once stood, and they asked for help in stopping the devastation."
– Natural Resources Defense Council, NRDC.org; August 2015

Canada's boreal forest and tar sands mining

Many people have no idea what is going on in the boreal forest areas of Canada. But, there is massive and extreme destruction taking place at the hands of the petroleum industry. Some people consider this to be the worst environmental disaster in the history of the planet. (I consider the entire fossil fuel industry combined with the meat industry to be that.)

"Tar sands oil – even the name sounds bad.

And it is bad. In fact, oil from tar sands is one of the most destructive, carbon-intensive and toxic fuels on the planet. Producing it releases three times as much greenhouse gas pollution as conventional crude oil does. Tar sands oil comes from a solid mass that must be extracted via energy intensive steam injection or destructive strip mining techniques that completely destroy ecosystems, put wildlife at risk, and defile large areas of land. Finally, when transported by pipeline or rail, it puts communities, wildlife, and water supplies in danger of toxic spills that are nearly impossible to clean up.

In early 2014, the Center for Biological Diversity launched an aggressive, coordinated effort to stop reckless tar sands extraction both in the United States and Canada. Our efforts have targeted cross-border pipeline proposals like Keystone XL and Alberta Clipper, crude-by-rail expansion, and burgeoning domestic tar sands projects.

Tar sands and other destructive fossil fuel projects pose a huge risk not only to people and wildlife, but to the future of a livable planet."
– Center for Biological Diversity, *No Tar Sands*; BiologicalDiversity.org

Indigenous people and tar sands mining

"Oil companies have replaced Indigenous people's traditional lands with mines that cover an area bigger than New York City, stripping away boreal forest and wetlands and rerouting waterways.

Oil and gas companies like ExxonMobil and Canadian giant Suncor have transformed Alberta's tar sands – also called oil sands – into one of the world's largest industrial developments. They have built sprawling waste ponds that leach heavy metals into groundwater, and processing plants that spew nitrogen and sulfur dioxide into the air, sending a sour stench for miles.

The sands pump out more than 3 million barrels of oil per day, helping make Canada the world's fourth-largest oil producer and the top exporter of crude to the United States. The economic benefits are significant: Oil is the nation's top export, and the mining and energy segment as a whole accounts for nearly a quarter of Alberta's provincial economy. But the company's energy-hungry extraction has also made the oil and gas sector Canada's largest source of greenhouse gas emissions. And despite the extreme environmental costs, and the growing need for countries to shift away from fossil fuels, the mines continue to expand, digging up nearly 500 Olympic swimming pools-worth of earth every day."

– Nocholas Kusnetz, *Canada's Tar Sands: Destruction So Vast and Deep It Challenges the Existence of Land and People*; InsideClimateNews.org, Nov. 2021

Billions pumped into tar sands

Tar sands have become a major source of petroleum, and billions of dollars are being pumped into the tar sands industry of not only Canada, but other areas of Earth. The governments of many countries help finance the fossil fuels industries, including the mining of tar sands.

Poisoning the Athabasca River Basin

The Athabasca River Basin of Canada has been intensely impacted by oil sands mining and the processing of the oil to produce fuel, which involves removing the coke and sulfur. Within a short number of years, the people living there witnessed thousands of acres of trees in the surrounding boreal forest being clearcut by the mining companies. The land, air, and water have been becoming more polluted. Wild animals are dying, found with cancerous growths, and are becoming scarce where they once flourished.

Songbirds dying and communities poisoned

"Our continent's greatest songbird nursery is in grave danger as oil companies violently strip-mine tar sands – the world's dirtiest oil – in the heart of Canada's Boreal forest. The resulting devastation could ultimately claim some 160 million migratory birds – including many of the songbirds we love seeing and hearing every summer.

Now [in 2011], a new oil industry plan would drive even more destruction of the Boreal while threatening the U.S. with environmental havoc. Their proposed Keystone XL pipeline [perhaps permanently blocked in 2021] would transport raw, toxic tar sands oil right through the American heartland – from Montana to Texas – endangering fresh water for millions of American families.

This continent-wide disaster is designed with one goal in mind: boosting oil company profits. Everyone else loses. But the U.S. State Department is rushing toward approval of this fiasco – unless we mobilize swiftly to stop them.

All you need to do is look out your window to grasp how crucial the Boreal is. Four out of ten 'backyard birds' in America migrate to the Boreal's wetlands to nest and rear their young. Warblers, goldeneyes, flycatchers, grosbeaks, and many more.

Tar sands mining is obliterating their habitat at a frenetic pace, instantly transforming majestic ancient forests into lifeless open pits. Four tons of wilderness must be dug up to yield a single barrel of tar sands oil.

The toll on wildlife is unimaginable. But songbirds may not be the pipeline's only victims. Native communities downstream from the tar sands have seen their cancer rates – which may be linked to mining – go up by 30 percent.

And Americans could be next. The 2,000-mile pipeline would carry hot, toxic and explosive 'diluted bitumen' – raw tar sands oil – from Alberta, Canada all the way to refineries in Texas, passing right through the vast Ogallala Aquifer, which supplies drinking water to millions of Americans in the heartland."
– Natural Resources Defense Council, NRDC.org; Aug. 2011

Keystone XL tar sands pipeline for the filthiest fuel

"The takedown of the notorious Keystone XL (KXL) tar sands pipeline will go down as one of this generation's most monumental environmental victories. After more than 10 years of tenacious protests, drown-out legal battles, and flip-flopping executive orders spanning three presidential administrations, the Keystone XL pipeline is now gone for good. The project's corporate backer – the Canadian energy infrastructure company TC Energy – officially abandoned the project in June 2021 following President Joe Biden's denial of a key permit on his first day in office. But the path to victory wasn't always clear. Many had hoped that the disastrous project was finally done for in November 2015, when the Obama administration vetoed the pipeline – acknowledging its pervasive threats to climate, ecosystems, drinking water sources, and public health. But immediately after taking office, President Trump brought the zombie project back to life, along with the legal battles against it. By the time President Biden took office in 2021, ready to fulfill his campaign promise to revoke the cross-border permit, the dirty energy pipeline had become one of the foremost controversies of our time.

Energy lobbyists claimed developing tar sands would protect our national energy security, and bring down U.S. fuel prices. But environmental reviews by both the Obama and Trump administration concluded that the Keystone XL pipeline would not have lowered gasoline prices. NRDC and its partners also found the majority of Keystone XL oil would have been sent to markets overseas.

Tar sands oil is thicker, more acidic, and more corrosive than lighter conventional crude, and this ups the likelihood that a pipeline carrying it will leak. Indeed, one study found that between 2007 and 2020, pipelines moving tar sands oil in Midwestern states spilled three times more per mile than the U.S. national average for pipelines carrying conventional crude.

Complicating matters, leaks can be difficult to detect. And when tar sands oil does spill, it's more difficult to clean up than conventional crude because it immediately sinks to the bottom of the waterway. People and wildlife coming into contact with tar sands oil are exposed to toxic chemicals, and rivers and wetland environments are at particular risk from a spill. Keystone XL would have crossed agriculturally important and environmentally sensitive areas, including hundreds of rivers, streams, aquifers, and water bodies. One was Nebraska's Ogallala Aquifer, which provides drinking water for millions, as well as 30 percent of America's irrigation water."

– Natural Resources Defense Council, *What is the Keystone XL Pipeline?*; NRDC.org, March 2022

Tar sands: the master plan

They Keystone XL pipeline was to run from Alberta, Canada, across Saskatchewan, into Manitoba, and down through North and South Dakota, to Nebraska, then needed to be built through Kansas and linked to a prebuilt pipeline from Oklahoma to Port Arthur, Texas. It was to speed up the transport of oil to refineries on the Gulf of Mexico. The products would then be exported.

Because of growing environmental concerns, fuel prices, and what happened with the Keystone XL pipeline, tar sands mining slowed, and Teck Resources canceled plans to create the largest Canadian tar sands mining operation in the world.

Pipeline networks continue the threat

Unfortunately, stopping the Keystone XL pipeline does not stop the mining of tar sands in Canada. There are other networks of pipelines – and also trains – transporting the filthy oil to refineries in the U.S. Part of that includes about three million miles of crude oil pipelines in Alaska and the lower 48 states.

All facets of the tar sands industry pose a threat to the environment. Its mines are a blight on Canada's boreal, where mining operations dig up and flatten forests to access the oil below, destroying wildlife habitat and one of the world's largest carbon sinks. The mining depletes and pollutes freshwater resources, creates massive ponds of toxic waste, and threatens the health and livelihood of the First Nations people who live near them. Refining the sticky black gunk produces piles of petroleum coke, a hazardous, coal-like by-product. What's more, the whole process of getting the oil out and making it usable creates three to four times the carbon pollution of conventional crude extraction and processing."

– Natural Resources Defense Council, *What is the Keystone XL Pipeline?*; NRDC.org; March 2022

Tar sands mining, ruined land, poisoned people, trashed Nature

A couple of decades ago, people in the boreal forest region of Canada being mined for tar sands drank the local water. By the 2010, they largely switched to drinking bottled water. They once caught and ate local fish. Then they became afraid to drink the local water or to eat the local fish. This is because of the heavy metals and chemical pollution from the petroleum industry. They used to enjoy clean air, but now have become accustomed to air pollution and the putrid smell of petroleum and chemicals. Respiratory issues have increased. Skin rashes have become common. Their cancer rates have risen. And their dependence on imported foods and products has increased. To survive, many of the people have found jobs working for the very same companies destroying their environment and culture, and killing the local wildlife. Their economy has become dependent on the monsters that are corporate greed and North America's demand for petroleum.

Encroaching mining industry poisoning wildlife

As their lives have been changed by the encroaching mining industry, the surrounding landscape has turned into poisoned ponds, rivers, and lakes made toxic by the mining industry as it takes approximately three barrels of water to obtain one barrel of tar sand oil. The landscape is also devoid of trees, but utility lines for electricity have sprung up. Where there were once pristine forests where wildlife thrived among streams, and meadows, there are thousands of acres of mining destruction void of the former wildlife. Where moose roamed, there are few or no large animals. Where there were no fences, there are now lines of fences topped by barbed wire to stop people and large animals from entering into toxic lakes. But birds and small creatures are not deterred by fences, and many end up dead from drinking or frolicking in the poisoned water. It is common to find dead waterfowl and small animals poisoned by the toxic water.

Petroleum, tar sands, and the Athabasca region

Tar naturally seeps from the ground in the Athabasca region. Cree used it as a building material and to waterproof their canoes. As people began to investigate the amount of petroleum contained in the region, it was found that the largest petroleum reserve outside of Saudi Arabia might exist in the tar sands of the Athabasca River Basin.

Because the technology didn't exist in a way that would be profitable until the late 1960s, getting petroleum out of tar sands had not been a process many people had been involved in.

In the U.S., where people use petroleum from tar sands every day, a majority of the people don't know about it, and many who do know about it lack an understanding of what tar sands mining entails.

I asked friends what they knew about tar sands. Most of them either didn't know about tar sands, or they had a very limited understanding about tar sand mining, processing, and pollution.

Tar Sands Gigaproject's massive destruction of Nature

"The Tar Sands 'Gigaproject' is the largest industrial project in human history and likely also the most destructive. The tar sands mining procedure releases at least three times the CO_2 emissions as regular oil production and is slated to become the single largest industrial contributor in North America to climate change.

The tar sands are already slated to be the cause of up to the second fastest rate of deforestation on the planet behind the Amazon Rainforest Basin. Currently approved projects will see 3 million barrels of tar sands mock crude produced daily by 2018; for each barrel of oil up to five barrels of water are used.

Human health in many communities has seriously taken a turn for the worse with many causes alleged to be from tar sands production.

Tar sands production has led to many serious social issues throughout Alberta, from housing crises to the vast expansion of temporary foreign worker programs that racialize and exploit so-called non-citizens.

Infrastructure from pipelines to refineries to super tanker oil traffic on the seas crosses the continent in all directions to all three major oceans and the Gulf of Mexico.

The mock oil produced primarily is consumed in the United States and helps to subsidize continued wars of aggression against other oil producing nations such as Iraq, Venezuela, and Iran."
– OilSandsTruth.org

Tar sands feed U.S. petroleum addiction

Driven by the U.S. addiction to petroleum, Canada became the number one supplier of foreign oil to the U.S. People consider this to be a good thing, because, they reason, the petroleum is coming from their own continent, and not from war-torn countries (which are often war-torn because of foreign countries interested in extracting the resources those countries contain).

In addition to engine fuel, kerosene and butane also are produced from tar sands.

Tar sands mining has only turned into a big deal since the September 11, 2001 airplane disasters in the U.S. Part of the reason the tar sands industry in Canada has expanded so quickly is because of recent technology, it is a domestic oil in Canada, and because, in the U.S., it is an oil from a friendly neighbor, and not from a politically tender and distant region. Because of this, in recent decades, tens of billions of dollars have been pouring into the Canadian tar sands mining operations, including pipelines and roads to bring the tar sands into the U.S.

The environment, the local and migrating wildlife, and the indigenous peoples of the region don't seem to matter, as long as the petroleum companies and their stockholders make money. Greed.

Tar sands service companies

Through my work with homeless people I met a woman who spent time among the tar sands mining workers. She told me stories of how there are a disproportionate number of men in the tar sands region of Canada, and women travel there to provide "services" for the miners. There has also been an increase in the illicit drug trade, and alcohol sales in the region as workers seek out various ways to escape the boredom of a mining town. The cost of everything has increased, from homes to food, services, and supplies, resulting in an economy long-term residents and Native Peoples had never experienced.

Gaping wound in Earth

While the petroleum companies make a killing through Canadian tar sands, what they are doing is creating a gaping wound in Earth that is helping to kill the planet.

The tar sands are spread out over a large area, and are not very deep. This means large areas are cleared of trees and soil to get to the sands saturated and streaked with petroleum. The result is sprawling and growing destruction of wildlife habitat, grassy meadows, and forests that produce oxygen, clean water, absorb and sequester greenhouse gases, and protect ground water, ponds, streams, rivers, and lakes – and oceans.

Tar sands water pollution

Because about three times as much water is used to produce one barrel of oil from tar sands, enormous amounts of water are polluted. The production of engine fuel from tar sands mining is about 50% more polluting than the same amount of fuel from drilled petroleum.

In other words, tar sands mining and refining processes are filthy. More pollution is created when the fuel is burned, spreading it far and wide into the atmosphere, helping to cause ocean acidification.

Corporate media and petroleum industry

With as much damage as tar sands mining is doing to the planet, it seems everyone would know about it, and it would be reported in the news for being so ruinous. However, that sort of commentary can't make it into the corporate-controlled media, which makes its money by selling advertising time to the fossil fuels, and the car and truck industries.

Global damage from tar sands mining

Tar sands mining decreases wildlife diversity, destroys and fragments wildlands, reroutes or eliminates streams and rivers, leaves behind toxic ponds, and causes global air, soil, and water pollution.

"Tar sands oil contains, among other toxic metals, 11 times more sulfur and nickel, six times more nitrogen, and five times more lead than conventional crude oil. These cause health issues in humans and wildlife.

Tar sands wastewater contains benzene, cyanide, phenols, toluene, arsenic, copper, sulphate, and chloride. Heavy metals and polycyclic aromatic hydrocarbons released in tar sands refining have been linked to pre-natal brain damage, and to learning disabilities. Communities downstream from the mines experience high rates of bile duct cancer.

Nitrogen oxides, along with volatile organic compounds released in tar sands refining are the principal causes of smog and ground-level ozone. Exposure to nitrogen oxides is a direct cause of asthma, emphysema, and other lung diseases.

There are 1,900 miles of Canadian oil pipelines in and around the Great Lakes watershed, the water source for 25 million people."
– *Toxic Tar Sands: Profiles from the Front Lines*; SierraClub.org

Pollution from tar sands mining and processing, and burning the fuel, ends up in the snow and rain, on vegetation, and in soil, aquifers, wetlands, marshes, streams, rivers, lakes, oceans, wildlife, and you.

Humans accumulate environmental toxins, especially if they eat high on the food chain, such as meat and dairy. The pollutants increase the risk of hormonal imbalances, learning disabilities, nerve disorders, autoimmune issues, cancers, birth defects, and miscarriages.

Ecocide disaster of tar sands mining

Tar sands mining is an obscene ecocide disaster for Nature. Pipelines continue being built from mining areas to refining and shipping facilities.

Great Bear Rainforest

A pipeline had been planned through British Columbia's Great Bear Rainforest to the West coast for export to Asia. This was stopped in 2016. Oil tankers would have had to make it through a network of islands to dock in an ecologically pristine area. If a tanker had an accident, the crude in the water would have been ecologically devastating. (Search: Prince William Sound disaster, Exxon Valdez.)

Coastal First Nations Great Bear Initiative

"Coastal First Nations have upheld a ban on oil tankers carrying crude oil through our Traditional Territories. The Coastal First Nations Great Bear Initiative has led wide-spread opposition to the Enbridge Northern Gateway Project, and called for a moratorium on all oil tanker traffic on the north Pacific Coast.

In November 2016, the federal government announced two decisions: Canada quashed the Enbridge Northern Gateway Project, and its proposed crude oil pipeline and supertankers, saying 'The Great Bear Rainforests is no place for a pipeline.'

This is a double victory for coastal First Nations who depend on a healthy ocean for our food, livelihoods, and cultural ways of life. Our communities have stood together in making choices that will safeguard our air, lands, rivers, salmon, and communities for the future.

The 1,170-kilometer Enbridge pipeline proposed to transport more than a half million barrels of bitumen a day from Alberta to Kitimat on the British Columbia north coast. More than 220 supertankers a year were to travel through our coastal waters, carrying crude oil for export to overseas markets.

The pipeline would have crossed more than 1,000 streams and rivers – over 800 of them in BCs Skeena and upper Fraser watershed. They would have traversed earthquake and avalanche-prone regions before reaching the fragile ecosystems of the West Coast."

– Coastal First Nations: Great Bear Initiative, Protecting Our Coast: Tanker Ban and Death of Northern Gateway Pipeline; CoastalFirstNations.ca

Pipeline ruptures

Pipelines often rupture, and if any are built, huge environmental disasters happen amongst more wildlands. More forest is demolished, more wildlife are killed and lose habitat, and more land is damaged by the construction and maintenance of pipelines – which require tremendous resources to build, including steel, concrete, and fuel.

Millions of barrels of oil have already been spilled as a result of a number of other pipeline failures in the U.S.

An oil spill among the forests, mountain valleys, and rivers of Canada, and/or in the temperate rainforests of the coast and/or along the coastal waters of British Columbia would do great harm to wildlife, and would damage tourism and other industries.

Tar sands expansion into Venezuela and Madagascar

As the process of tar sands mining is expanding into other countries, including Venezuela and Madagascar, there had been big plans to mine more and more tar sands in the Colorado watershed region of the U.S., including Utah.

Environmental groups are working to educate people about the environmental damage being done by the tar sands industry. Among those groups are the Natural Resources Defense Council, the Environmental Defense Council, GreenPeace, Earth First!, The Sierra Club, and Earth Island Institute. (For more information, access: OilSandsTruth.org.)

The Ogallala Aquifer

The KeystoneXL tar sands fuel line planned from Canada to Texas would have polluted the Ogallala Aquifer, which is a major source of drinking and industry water for cities, towns, and farms in several states, including Wyoming, South Dakota, Nebraska, Texas, New Mexico, and Oklahoma.

Property owners along the route had been pressured into selling to the petroleum company, TransCanada. Some had been told the company would go to court to use the power of eminent domain to gain access to properties.

Circulation of water

"When you think about our planet, it is one huge living body because of the water that flows through it. And, across the land masses you can have veins of water and arteries of water flowing. This flows into our ocean, which is the heart of our Earth, and it exhales, evaporates water and puts water back into the hydrologic cycle, which then again goes up to the mountain tops and then again they form veins and arteries. So, our planet is just one, huge living body with water constantly circulating through it, and it is the water that gives it life. We too are like the surface of our planet, 70% water and 30% solid. We too have a heart. We have 60 thousand miles of veins and arteries. And just like the Earth recycles the water, we have a water cycle within us."

– William E. Marks, author of *The Holy Order of Water*, speaking in the documentary, *Flow*; FlowTheFilm.com.

Depletion and poisoning of Ogallala Aquifer

When you mention the Ogallala Aquifer to people, most have no idea what it is. Yet, most of them are consuming food grown on farms dependent on the Ogallala Aquifer. What they also don't know is the Ogallala Aquifer has been depleted, and much of it has become contaminated by chemicals used in a variety of industries, including chemicals spewed by airplanes and motor vehicles, chemicals used by the farming industry, and chemicals used in the natural gas, petroleum, coal, and mining industries.

Why would the damage done to the Ogallala Aquifer matter so much?

Less than 3% of the water on the planet is fresh water, and one of the largest sources of freshwater in North America is the Ogallala Aquifer, which supplies everything from drinking water to water used to irrigate crops, wash clothes, clean medical equipment, water lawns, and feed animals.

Largest use of water from Ogallala Aquifer

The largest use of the water from the Ogallala Aquifer is related to animal farming, especially to grow massive fields of monocropped feed for billions of farmed animals.

As mentioned elsewhere, most of the food grown in North America – and other continents – is fed to farmed animals, not to people. In turn, the people eat the animals. It takes at least 75% less water to support a vegan diet than it does to support the diet of the average meat eating American. The chief cause of the depletion of the Ogallala Aquifer has to do with meat consumption.

Cotton crop farming and water

Another water-intensive crop is cotton. The cotton industry also uses a tremendous amount of farming chemicals, including fertilizers and pesticides.

A much safer crop to grow is hemp, which provides fiber that can be made into more varieties of fabric than cotton, uses less water to grow, sequesters more greenhouse gasses per acre than trees, is safer for the soil and wildlife, and doesn't require the chemicals to grow to get an equal amount of fiber than what cotton provides.

Natural gas is a fossil fuel

As the tar sands industry continues in Canada, with plans to tear into and poison U.S. land to gain access to more tar sands, the natural gas industry is already ripping up land throughout the U.S., From New England to Pennsylvania to the Midwest to Texas, Montana, Wyoming, and other states. Roadways are being built, old roadways are being widened, pristine land is being cut into, wildlife is being damaged, species reduction is happening, water is being poisoned, soil is being saturated with a variety of harmful chemicals, gasses are being spewed into the atmosphere, networks of pipelines and electrical lines with accompanying service roads are being built through the former wildlands, and company vehicles are compacting soil and destroying delicate ecosystems that support small forms of life – which support larger wild creatures.

The natural gas industry is reaping the financial rewards for their environmental destruction as they are selling more and more natural gas to satisfy market demand. All this while increasing the environmental destruction, damaging wildlife, and increasing human illness. As long as the fossil fuels industries make more money, it's all acceptable. I guess.

Fracking filthy

While many people consider natural gas to be a safe fuel, they are not dealing with reality.

If you think coal causes massive environmental damage, which it does, consider that natural gas is even worse, which it is.

Hundreds of thousands of natural gas wells have been drilled across the North American continent, and each of those wells is causing harm to the land, water, air, and wildlife.

To understand what "fracking" is, and how terrible it is for the environment, for wildlife, and for us, access: GaslandTheMovie.com, and watch the documentary. (Search: Fracking environmental damage.)

"Basic chemistry leaves us in little doubt that our burning of fossil fuels is changing the acidity of our oceans. And the rate of change we are seeing to the ocean's chemistry is a hundred times faster than has happened for millions of years.

Failure to (cut carbon dioxide emissions) may mean there is no place in the oceans of the future for many species and ecosystems that we know today."

– John Raven, oceanic expert with Britain's Royal Society, which reported the oceans were absorbing one ton of carbon dioxide (the primary greenhouse gas) per person per year, and are running out of the capacity to absorb it; June 2005

Demand for fossil fuels and energy

Unfortunately, the demand for petroleum, natural gas, and coal keeps increasing, not decreasing. Car culture keeps spreading, and what are chief components of the fossil fuels industry market – the meat and affiliate industries, and the consumption of meat – keep increasing. Companies working to satisfy the demand for fossil fuels keep destroying the planet. And they keep making billions of dollars, which allows them to afford to manipulate governments, put forth huge public relations campaigns, and cover the legal expenses of conducting such environmentally destructive operations and forming government policies to help them to do so.

Offshore oil drilling and US President approvals

When Obama appointed Ken Salazar as Secretary of the Interior, one of the first things Salazar did was to place a record 53 million offshore acres up for lease to various oil and gas companies. This was done under the administration of Obama, a man who promised change – but whose administration was saturated with petroleum industry buddies. Of course, it didn't help their image when the Deepwater Horizon rig blew up in April, 2010, spilling an unknown amount of raw petroleum into the Gulf of Mexico, killing unknown millions of sea creatures. In 2023, those toxins continue as pollution and toxins bioaccumulating in wildlife.

While Trump did accept enormous amounts of campaign donations from the fossil fuels industries – as did other candidates, including Hillary Clinton – and made efforts to restart the XL pipeline, and other disastrous fossil fuel projects, surprisingly, as of 2023, President Biden had approved more fossil fuel drilling then Trump. It's so sickening.

Politicians catering to fossil fuels industries

Numerous American presidential administrations have been connected to the petroleum industry. There is huge money in fossil fuels and the industry spends grandly in Washington on lobbyists, and on local and national campaigns, and other political donations. Many politicians and/or their families have connections to the fossil fuels industries – including by owning land where fossil fuels are extracted.

Much of U.S. government policy has been formed by and to satisfy the fossil fuels industries – including in what amounts to wars in other countries to gain and secure access to resources.

Energy industries and politicians

U.S. Presidents and candidates for the office continue their grotesque distortions of environmental friendliness talk when speaking of the energy industries. They speak of building more nuclear power plants (always a bad idea) as a way of creating sustainable energy (calling nuclear energy a sustainable fuel is a lie), speak of clean coal (there is no such thing), toss around talk of capping carbon emissions (it sounds nice, but what are they really doing – but expanding and helping to fund the fossil fuels industries), continue subsidies to the corn industry (to create starch ethanol that is not as good as cellulosic ethanol), and have not done much to really develop truly alternative fuels, such as cellulosic ethanol (using landscape clippings and hemp), have not put an end to offshore drilling, have not done anything to build one single monorail in the cities (monorails should be built and bike lanes established in all of the major cities), have not worked to give significantly impressive tax credits to people for installing bird-safe wind turbines and solar energy panels on their homes and businesses, and have not allowed for or invested in the development of an industrial hemp industry within the U.S. (Hemp is not a drug, and can't get you high, but can supply fabric, paper, fuel, and building materials more environmentally safe than those products made from cotton, tree pulp, tree wood, fossil fuels, plastics, and concrete). Instead, the administrations are stacked with friends of Monsanto, friends of petroleum, friends of coal and natural gas, friends of the chemical industries, friends of the corrupt banking and stock market industries, friends of the pharmaceutical industry, friends of the medical industry, friends of the animal farming industry, friends of the feed crop industries, and friends of the billionaire boys club.

Fossil fuels industry control over government

How the Deepwater Horizon gusher tragedy in the Gulf of Mexico was handled is an example of the control the petroleum industry has over the U.S. government. Not only should such deep water wells be banned, they should have never been permitted.

Journalists arrested for reporting on Deepwater Horizon

The petroleum company and its government minions should not have been permitted to order the arrest of journalists and charge them with felonies and up to $40,000 in fines when the journalists were trying to report on the gusher and the subsequent environmental damage. Yet, that is what was being done, with the Coast Guard, private security firms, and local police working off duty for the petroleum company. Even the FAA placed a restricted fly zone over the Gulf areas and over beaches impacted by the spill – and those restrictions were largely put into place to prevent the accurate reporting of the massive tragedy.

The petroleum industry should not have been permitted to spray and apply millions of gallons of highly toxic petroleum-based chemical dispersants over and onto the Gulf, which greatly added to the tragedy, killing all sorts of wildlife, including sharks, dolphin, whales, turtles, manatees, fish, crustaceans, coral, and birds.

The damage caused by the dispersants continues on into 2023, as the petrochemical poisons remain in the water and some wash onto the shores and into coastal marshes, swamps, and waterways.

Meanwhile, to put it plainly, the demand for fossil fuels wouldn't be so high, and these companies wouldn't be so in control if more effort were made to get rid of the demand for fossil fuels.

Going sustainable

The people of the U.S. need to both drastically decrease their use of fossil fuels, and increase their self-reliance in a sustainable way, such as by joining bike culture, using compostable products, growing some of their food using organic gardening practices, and eliminating toxic chemicals from their life. (See the documentary: *The Need to Grow*)

Ruining the planet is ruining themselves

Most people in the U.S., and other industrialized countries, continue to rely on supermarkets, restaurants, snack shops, and other corporate food outlets to supply every calorie of their food. They are also caught in the thought pattern they need to look to political and business leaders to take the lead. Instead, they can make changes in their own life, and stop giving power to politicians and corporations, which largely work in their own interests of making money and becoming more powerful, and holding onto the power (which is truly their illusion – as ruining the planet is ruining themselves, and the lives of their families).

"If Louisianans have been willing to sacrifice their unique landscapes for oil and gas, the rest of us put them up to it. We do so every time we put the key in the ignition or enjoy a 15-minute hot shower fueled by natural gas."
– Jason Mark, *We Are All Louisianans; Earth Island Journal*, 2010

Levels of destruction

Meanwhile, the environment continues to be degraded by the fuels industries. Mountaintops continue to be demolished by coal mining companies. The natural gas industry keeps expanding and is polluting the land with horrible chemicals damaging to wildlife and humanity. The petroleum industry keeps mining and drilling. The nuclear energy industry, the uranium mining it does, radioactive waste it leaves behind, and the potential for meltdowns remain threats to life (not that the fossil fuel industry isn't), and continues to expand. And plastics made from fossil fuels continue to be more commonly used, and end up as trash.

Getting off fossil fuels

If people want to make a change in the environment, they need to greatly reduce their use of fossil fuels, including through converting away from petroleum-fueled engines, and cutting out their dependence on coal and natural gas. There needs to be a tremendous reduction of meat and dairy, while consuming fewer imported foods, and relying more on locally grown foods – including by growing organic food home and community gardens and supporting local organic produce farmers. Alternative fuels, including cellulosic ethanol, plant oil diesel, solar, hemp, and wildlife-safe wind energy are options.

War on global warming

Taking some of the trillions of dollars currently going into the war and weaponry industries, and using it to transform cities and towns to be more sustainable would be a very good thing. Instead of a war on foreign regimes and to gain and secure access to "natural resources," the real war should be fought against global warming, and it should be fought with home gardens, with organic farming, with solar electric generating plants placed on sprawling rooftops, with low-profile, wildlife safe wind turbines, with energy efficient technology, with an industrial hemp industry, with a spread of bike culture, with a tremendously vast reduction in plastic and pharmaceutical pollution, and with more people following locally-sourced plant-based diets.

The CO_2 fiasco with calcium carbonate life forms

"Even if you are a hard-bitten global warming skeptic – and, I talk to that community fairly often – you cannot deny the simple physics of CO_2 dissolving in the ocean.

We are pumping out lots of CO_2 into the atmosphere from fossil fuels, and from cement production. Right now about a third of that carbon dioxide is dissolving straight into the sea. And, as it does so, it makes the ocean more acidic. You cannot argue with that. That is what is happening right now. And it is a very different issue than the global warming issue. It has many consequences. There are the consequences for carbonate organisms. There are many organisms that build their shells out of calcium carbonate. Plants and animals, both.

The main framework material of coral reefs is calcium carbonate. That material is more soluble in acidic fluid. So, one of the things we are seeing is organisms are having to spend more metabolic energy to build and maintain their shells. At some point as this transient, as this CO_2 uptake in the ocean continues, that material is going to start to dissolve, and, on coral reefs, where some of the main framework organisms disappear, we will see a major loss of marine biodiversity.

But, it is not just the carbonate producers that are affected. There are many physiological processes influenced by the acidity of the oceans – many reactions involving enzymes and proteins are sensitive to the acid content of the oceans. So, all of these things – greater metabolic demands, reduced reproductive success, changes in respiration and metabolism – these are things we have good physiological reasons to expect to see stress caused by transient.

We have figured out some pretty interesting ways to track CO_2 levels in the atmosphere, going back millions of years. We used to do it with ice cores. But, in this case, we're going back 20 million years. And we take samples of the sediment and it tells us the CO_2 level of the ocean, and therefore, the CO_2 level of the atmosphere. And, here is the thing: You have to go back about 15 million years to find a time when CO_2 levels were about what they are today. You have to go back about 30 million years to find a time when CO_2 levels were double what they are today. Now, what that means is that all of the organisms that live in the sea have evolved in this chemo stated ocean, with CO_2 levels lower than they are today. That's the reason that they are not able to respond or adapt to this rapid acidification that is going on right now."
– Rob Dunbar, oceanographer, biologist

People need to learn what has been going on with Earth, and the destruction of it, and understand the way they are living is helping to destroy what they need to survive: Nature.

Salt marsh destruction

"For at least 75 years, the unique ecosystem of Southern Louisiana – site of 40 percent of the United States' saltwater marshes – has been disappearing. Each day, an area the size of a football field goes under

water. In the last half-century, Louisiana's coastal wetlands have vanished at an average rate of 34 square miles a year. One of the main drivers of this loss is the oil and gas industry, which, since the 1930s, has dug some 8,000 miles of canals and pipelines through the marshes, allowing seawater to intrude inland and destroy fragile coastal grasses. Before the oil disaster made Louisiana's wetlands a national emergency, the place was already suffering from chronic distress."
– Jason Mark, *We Are All Louisianans; Earth Island Journal*, 2010

Niger Delta: a petroleum nightmare assaulting Nature

While much of the attention on petroleum disasters has been on the Gulf of Mexico, and when incidents like when the Exxon Valdez oil tanker ran aground and spilled 11 million gallons of crude oil into Alaska's Prince William Sound, there has been a continuing and growing disaster taking place in Africa's Niger Delta.

Since the 1970s, the Niger Delta has been getting poisoned with millions of barrels of crude oil. Part of the problem is the lack of regulations limiting what can and can't be done in the quest for extracting petroleum from the lands of Nigeria. It is also a situation driven by companies intent on exploiting the lack of standards. Every year, millions of gallons of crude oil ends up in the Niger Delta.

The Niger Delta is 7,700 square miles of a 27,000 square mile delicate wetland and marine ecosystem that is home to hundreds of rare and endangered plants and animals. It is an important region for migrating birds and mammals. It is one of the largest deltas in the world, and is the largest wetland region of the African continent.

"A new oil spill at a Shell facility in Nigeria has contaminated farmland and a river, upending livelihoods in the fishing and farming communities in part of the Niger Delta, which has long endured environmental pollution caused by the petroleum industry.

The National Oil Spill Detection and Response Agency told the Associated Press that the spill came from the Trans-Niger Pipeline operated by Shell that crosses through communities in the Eleme area of Ogoniland, a region where the London-based energy giant has faced decades-long local pushback to its oil exploration."
– Taiwo Adebayo, *Oil spill from Shell pipeline fouls farms and a river in a long-polluted part of Nigeria*; Associated Press; June 2023

Niger Delta, from the land to the sea and air: all polluted

The spill described above polluted the Okulu River, which many communities depend on, and empties into the Atlantic. It impacted 300 fisheries, ruined soil of many farms, and impacted thousands of already struggling, poor, and marginalized people.

Not only are crude oil spills poisoning the wetlands of Niger, the petroleum companies also pollute the air by gas flaring, which is the leading cause of air pollution in the region, and releases millions of tons of carbon into the atmosphere.

In 2009, Royal Dutch Shell was responsible for spilling more than four million gallons of crude oil into the Niger Delta. Royal Dutch Shell is only one of many companies that have been drilling for oil in – and creating an ecological ecocide nightmare in – the Delta.

The Nigerian Justice League, Friends of the Earth International, The Yes Men, and Earth Island Institute are some of the organizations working to bring an end to the deplorable practices of oil companies drilling in the Niger Delta. So far, the efforts have not been successful. Big oil continues ruining wildlife. Oil spills continue. Additionally, vandalism of pipelines, and badly maintained pipelines have led to fires, explosions, and the deaths of people and a wide variety of wildlife.

Wherever wildlife is damaged, so too are human lives.

Unfortunately, the Niger Delta is not the only region of Africa being violated by the petroleum industry. Other tragic petroleum hotspots include Chad, where there are no environmental laws, a situation also being taken advantage of by the petroleum industry.

Everywhere you look Nature is being damaged

It seems everywhere you look in Nature, in one form or another, it is being damaged by the fossil fuels industries. It is all being driven by the human animals, which are the most destructive form of life on the planet, and one, by its common everyday choices, that is endangering its own survival, and the survival of most life forms currently sharing the planet.

As I write this book and research each topic, issue, and concern, I am becoming more and more aware of how much damage is being done to the planet, from pole to pole.

Wetland destruction

Wetlands throughout the world are often drained for "development" of housing tracts, warehouses, factories, office building campuses, airports, petroleum processing plants, military bases, prisons, cropland to grow food for farmed animals, and for factory farms and meat processing plants. It is as if thousands of years of Nature building up networks of wildlife is of no meaning compared to the need to make rich people wealthy, and supply people with what they don't need.

111

Airports and airlines and private jets

"We measured gases, particles, and ultrafine particles continuously for one month at a residence near the Logan International Airport, Boston. The residence was located under a flight trajectory of the most utilized runway configuration. We found that when the residence was downwind of the airport, the concentration of all gaseous and particulate pollutants were 1.1- to 4.8-fold higher than when the residence was not downwind of the airport. Controlling for runway usage and meteorology, the impact were highest during overhead landing operations: average particulate number concentration was 7-5-fold higher from overhead landings versus takeoffs on the closest runway. Infiltration of aviation-origin emissions resulted in indoor PNC that were comparable to ambient concentrations measured locally n roadways and near highways. In addition, ambient NO_2 concentrations at the residence exceeded those measured at regulatory monitoring sites in the area, including near-road monitors."

– Neelakshi Hudda, Liam Durant, Scott Fruin, John Durant, *Impacts of Aviation Emissions on Near-Airport Residential Air Quality*; American Chemical Society, Pubs.acs.org; July, 2020

Airport pollution and human disease

I had a friend who was working to make it as a singer songwriter in the Los Angeles music scene. To get used to playing his guitar and singing in front of audiences, he performed at various coffee houses and bars. To support himself during those years, he worked unloading luggage from jetliners on the tarmac of L.A. International Airport. Until his lungs started to hurt. He was diagnosed with lung cancer, went through aggressive medical treatments, but passed away.

"Airports are among the largest sources of air pollution in the United States. In fact, Los Angeles International Airport is the largest source of carbon monoxide in the state of California. Average airplane taxi time – the amount of time an airplane spends between the gate and runway – increased by 23 percent from 1995 to 2007. This increase in average congestion, combined with an increased number of flights, translates to an aggregate increase of over one million airplane hours per year spent idling on runways, leading to significantly higher levels of local ambient air pollution."

– National Bureau of Economic Research, *Airports, Air Pollution, and Health*; May 2012

"Aircraft engines are responsible for approximately 2% of global CO_2 emissions, while ground service equipment accounts for 9%.

On average, a single aircraft takeoff produces 9.5 kgs of nitrogen oxide emission, and airports in the Unites States generate over 16,000 tons annually.

The world's busiest 50 airports account for almost half of all global CO_2 from aviation. U.S. airports produced nearly 2.86 million tons of solid waste in 2018, which contributes to further air pollution levels as well as climate change."

– *Airport Statistics and Trends 2023*; GitNux.com, 2023

Private jet pollution

"Public scholars believe the explosion of private jet use has been unwelcome to the Earth and taxpayers.

One of every six flights the Federal Aviation Administration handles are flown by private jets.

Private jets emit at least 10 times more pollution than commercial plants per passenger. Emissions have increased by more than 23% as private jet use has risen by about a fifth since the COVID-19 pandemic began.

Billionaire Elon Musk's private jet activity was exposed last year, showcasing an astonishing amount of carbon pollution. The Institute for Policy Studies reported that Musk flew a total of 171 flights, or nearly one flight every two days in 2022, which consumed over 837,000 liters of jet fuel. The activity produced 2,112 tons of CO_2 emissions – equivalent to 132 times more than the average carbon footprint of a person."

– Channing Reid, *How Bad Are Private Jets For The Environment?*; SimpleFlying.com; July, 2023

"A typical private jet emits carbon at a rate of 4.9 kilograms per mile. A commercial plan emits about 85 grams per passenger per kilometer. In the United States, there are an estimated 18,000 private jets."

– FlyBitLux.com, Nov. 2022

Sprawling airports

One industry that has used up a tremendous amount of land on every continent is the airline industry.

International airports typically take up thousands of acres of land, and require massive support systems, including roads, vehicles, electricity, fuel, concrete, steel, and water. Each one is basically its own town producing a tremendous amount of pollution, and more pollution than a town.

One thing most common about international airports is they don't get smaller, they most often grow larger – taking up more land, using more resources, and producing more pollution.

Commercial airports, in one way or another, are also funded by tax dollars.

While private small planes and private jet flights have greatly increased over the past several years, by 2023, it has resulted in more than one-in-eight flights in U.S. air space being private planes. This has increased noise and pollution.

Besides noise and air pollution, airports pave over hundreds of acres of land, destroying natural landscapes, and killing or displacing varieties of wildlife.

New Mumbai airport decimating wildlife

Ten years ago in Mumbai, India, wetlands were drained, hundreds of acres of mangrove forests were decimated, and thousands of poor people were displaced to build an airport. Concerns of environmentalists, including those associated with the Conservation Action Trust, had been dismissed in the drive to satisfy corporations. The elimination of more mangrove forests to expand the airline industry adds to the problem of flooding, destroys more rare and endangered species, adds more pollution to the region, and degrades the conditions of the poor. The new airport is expected to open in 2030 – that is, if sufficient numbers of humanity still exist to finish and use the place.

Each generation of biologists

"Each succeeding generation of biologists has markedly different expectations of what is natural, because they study increasingly altered systems that bear less and less resemblance to the former, preexploitation versions."
– Paul Dayton; *Science;* 1998

Blue whales

I don't have to look so far to find examples of how humans are damaging the environment and wildlife. Within several miles of where I live, the largest species of animal that has ever lived, the blue whale, swims in the ocean. It is estimated there were once 250,000 to 350,000 blue whales. The species had nearly been hunted into extinction, until the International Whaling Commission banned blue whale hunting in 1966. As I am writing this paragraph in 2023, an estimated 10,000 blue whales survive.

"A century ago, at least 100,000 blue whales roamed the seas as the top predators in their food chain. Today, due to human whaling activities, irresponsible fishing gear disposal, and overall aquatic ecosystem deterioration, only 10 percent of the blue whale population has survived.

Blue whales have been moved from the top of their food chains to the top of the endangered species list. Human activity is the main cause of the drastic decline in blue whale populations, mostly due to fishing gear, vessel strikes, and whaling.

Noise pollution from human activities, like ship sonars, fracking, or most recently deep-sea mining have been found to interrupt whale communications. Due to this, whale communities may lose their migration routes and baby whales can become separated from their mothers, reducing their chance of survival.

– Gerardo Bandera, *The Blue Whale is Close to Extinction. How Many are Left?*; FairPlanet.org, March 2023

The more extinction of marine plant and animal species of the oceans, the less the oceans will sequester carbon, and the less oxygen the oceans will produce through the numerous forms of plant life existing in them.

One of the main enemies of blue whale is the international shipping industry. Many of those ships are transporting petroleum and new cars and trucks that run on that fuel, and also military equipment used to secure the access to foreign oil and other resources. Cruise and military ships also kill many whales. The ship crews don't always know they harmed or killed a blue whale because, unlike some sea life, blue whales sink when they die. Even when captains do see the blue whale in their path, they often can't avoid them – because ships can't quickly turn.

Electric vehicles and mining the oceans

How could this topic mix in with the ocean life issues? It does.

Currently, companies are looking into gaining more access to minerals at the bottoms of the oceans, including to manufacture more electric vehicles. Rocks at the bottom of some oceans contain cobalt, copper, manganese, and nickel. As electric cars, trucks, motorcycles, bikes, and other vehicles increase in popularity, as will the popularity of cell phones and other battery-operated devices, and the demand for other forms of electric energy storage, the rush to secure access to these minerals has increased, including minerals from the sea beds. This mining would add to the ocean noise that interferes with whale and other sea mammal activities, breeding, and migration – and lead to the deaths of more whales and sea mammals that spread nutrients and help the oceans stay alive. It would also disrupt the lives of unknown numbers of sea creatures, many of which still remain to be discovered.

Some companies are making batteries with fewer metals, and without certain things like cobalt. Some aim to make batteries free of toxic glues, which would make batteries easier to recycle. As inventions evolve and processes are developed, it becomes less likely that deep-sea mining will be as financially rewarding for investors. But the mining of nickel and other battery elements continues to destroy landscapes, and charging one car can use more electricity than hundreds of houses.

Cellulosic ethanol engine fuel, and plant-oil diesel

I wasn't for electric cars. I thought it would be better to have gasoline vehicles converted to run on cellulosic ethanol made from hemp, grown and processed locally in each region. The fields of hemp would absorb greenhouse gasses, and sequester them in the plants and soil. And converting diesel engines to run on hemp oil. This would require none of the mining that is otherwise done for electric vehicles. And it wouldn't mean there would be millions of EVs made, or millions of charging stations manufactured and installed (paving over even more land), requiring even more metals and other materials.

Improving the train systems, and putting solar- and magnetic-powered monorail systems in the major cities would also help to reduce vehicle use, and reduce all sorts of pollution.

Older cars dumped in poorer nations

Under current circumstances, older cars are exported to poorer countries, where cheaper and older cars are in high demand. This continues the use of combustible engines using petroleum, and shifts the pollution to those countries (where pollution standards are less likely to be strict, or enforced). It continues the pollution being spewed into the atmosphere, contributing to climate change and more environmental destruction. If anything, before being exported, the cars could be converted to run on cellulosic ethanol and plant oils, which means those countries could grow and produce fuels from easy to grow hemp. Trucks, busses, motorcycles, boats, tractors, lawn mowers, and other petroleum-fueled engines also need to be converted to pure plant fuels.

Transporting fossil fuels

As it is, fossil fuel use continues to damage marine life. Including in the shipping of enormous amounts of petroleum across the oceans. But also the pipelines that run over or under creeks, rivers, and lakes.

Michigan peninsula and the Enbridge Line 5 pipeline

A 643-mile "Line 5" pipeline built in 1953, and since added onto, runs from Wisconsin, across the Michigan peninsula, down into lower Michigan to Ontario. According to the Enbridge website (2023), every day the pipeline carries 540,000 barrels of light crude oil, light synthetic crude, and natural gas liquids. Most of the fuel is from the dirtiest and most environmentally destructive mining operation: Canadian tar sands. The pipeline remains a risk wherever it exists as fuel pipelines leak, and sometimes burn, spewing the toxic contents into water and on land, killing wildlife, and tainting soil to the point it can no longer support plant or animal life. Native Peoples of Wisconsin's Bad River Bend Reservation have been trying to shut down the pipeline, and have it removed. It crosses the reservation and the 44-mile Bad River. A pipeline rupture can do a tremendous amount of damage to the river and surrounding environment.

Some of the pipeline is of two, 20-inch diameter parallel pipelines beneath water in the gorgeous Straits of Mackinac between Lake Michigan and Lake Huron. Building this pipeline should never have been considered, approved, or built. Any areas of the Great Lakes are tremendously bad areas to have oil pipelines.

"Enbridge's Line 5 pipeline transports 22 million gallons of crude oil and natural gas liquids across 645 miles of countryside every day. This aging oil pipeline violates tribal treaty rights, and poses catastrophic risks to the drinking water for 40 million people, and **one-fifth of the world's surface freshwater**."
– Sierra Club Michigan Chapter; SierraClub.org, 2023

Anishinaabe legends and the Great Lakes

"In Anishinaabe legends and histories, Michelimackinac, at the confluence of three Great Lakes, is the center of the cosmos, and the birthplace of creation. Whitney Garvelle, president of the Bay Mills Indian Community, put it this way: 'If there were a Garden of Eden, that is what the Straits of Mackinec is to our people.'

Now the place of the great turtle is home to an aging pipeline known by an antiseptic name: Enbredge Line 5.

Since 1967, the pipeline has experienced numerous spills totaling well over one million gallons (as of 2017).

Another artery in the system, Line 78 (which travels from Duluth, to south of Chicago, and across Michigan), formerly known as Line 6b, is responsible for one of the largest inland tar sands oil spills in history.

In July 2010, near Michigan's Talmadge Creek – a tributary of the Kalamazoo River – a corroded segment of Line 6B split open. It was 17 hours before technicians in the Edmonton, Alberta, headquarters of Enbridge Pipeline Inc. learned of the rupture.

By the time they shut off the oil, according to the EPA, almost a million gallons of crude had spilled into Talmage Creek, which, swollen by recent rains, swept the oil into the Kalamazoo. Beth Wallace, a staffer at the Great Lakes Regional Center of the National Wildlife Federation, grew up in the vicinity of the spill. 'The river was completely saturated in oil. You couldn't even see water. It was just black, thick oil at the surface. The trees were saturated in oil. Everything.'

The Kalamazoo spill came as a shock to many Michiganders. To some, it also came as a revelation and an impetus to action.

Over the past decade, the Sierra Club helped lead a coalition of environmental activists, residents, and their attorneys, and all 12 federally recognized tribes in Michigan in opposition of Line 5, this became 'a citizens movement that politicians couldn't ignore,' said Michigan Chapter chair Anne Woiwode."

– Donovan Hohn, *A Battle for the Future of the Great Lakes*; SierraClub.org, March, 2023

Turtle Island and the Anishinabeck

"Look at the map of Turtle Island, which is how the Anishinabeck refer to North America. The heart of the turtle is here in the Great Lakes. Enbridge Line 5 is in the worst place on Earth for a pipeline. There is no way that anyone proposing a pipeline through this area today would ever get permission. Line 5 predated the Clean Water Act by almost 20 years. It was already here before most Americans started really paying attention to the environment."

– Kathie Brosemer, former environmental program manager for the Sault Tribe of Chippawa Indians

Why is what happens with this pipeline important? Legal issues associated with it can help determine court cases impacting what happens with other pipelines across the continent.

Straits of Mackinac are a tremendously sensitive area

"The Straits of Mackinac are one of the most iconic settings in the Great Lakes. They include hundreds of islands, and miles of shorelines rimmed with forests and wetlands. Scenic Mackinac Island in Lake Huron, a popular resort areas since the mid-1800s, is Michigan's top tourist destination.

The straits have long been spiritually important to Great Lakes tribes. Michigan acknowledges that the Chippewa and Ottawa peoples hold treaty-protected fishing rights that center on the Mackinac region.

In 2010, another Enbridge pipeline, Line 6b, ruptured near the Kalamazoo River in southern Michigan, spilling over 1 million gallons of heavy crude. Line 6b is part of a parallel route to Line 5, and the cleanup continues more than a decade later.

University of Michigan oceanographer David J. Schwab concluded that the Straits of Mackinac were the 'worst possible place' for a Great Lakes oil spill, because of high-speed currents that were unpredictable, and reversed frequently. Within 20 days of a spill, Schwab estimated, oil could be carried up to 50 miles from the site into Lakes Michigan and Huron, fouling drinking water intakes, beaches, and other critical areas."

– Mike Shriberg, Line 5 rupture would devastate Great Lakes. Should Engridge pipeline move because of the risk?; Detroit Free Press; Sept. 2023

All remaining areas of wild Nature are critical. All need to be protected. And more need to be restored, unpaved, and paradise replanted for local and migrating wildlife to flourish.

The risk of the badly-built Line 5

Line 5 crosses over more than 200 waterways, including creeks, rivers, wetlands, and lakes. The Sierra Club has reported the pipeline has had at least 34 known documented spills.

Studies have revealed Line 5 beneath the strait is lacking in protective coating and adequate anchors, which could lead to ruptures. The pipeline has been damaged by boat anchors, and parts of the pipeline have leaked more than a million gallons of natural gas liquids and petroleum oil.

What happened with the Line 6b can easily happen to the Line 5 – or to any other pipeline across the planet, devastating more land and marine life, tainting soil, aquifers, streams, rivers, lakes, wetlands, marshes, swamps, and oceans.

What's in the pipelines

The fossil fuels in the pipelines crossing through sensitive wildlands are refined into everything from engine fuel to propane, and even plastics. All so people can use cars, trucks, trains, planes, motorcycles, mopeds, race vehicles, boats, and yachts, take ocean cruises, have outside grills, use leaf blowers and lawn mowers, go "off roading" (destroying more land), and ride four wheelers, wave runners and snow mobiles. How much pleasure is taken in the destruction of Nature?

The above is only part of the problem about what is going on with the freshwater sources on the planet, and the life reliant on it. This, while the saltwater of the oceans continue to experience a wide variety of environmental concerns, including the dire situations and decreased populations of sea mammals, and other marine animals.

(Search: Permaculture. Also, learn about veganic agriculture.)

Killing sharks and other sea creatures

Unfortunately, blue whales are hardly the only local species in my region of the planet that are in danger.

Shark fins

In October 2011, California made it illegal to sell, trade, possess, or distribute shark fins. Sadly, shark fins and their sale and use are not outlawed in other states or countries. They are used in a "Chinese delicacy" called *shark fin soup*.

While I haven't seen many sharks, I know they are swimming along the California coast. One morning I did find a baby shark dead on the sand, and left it there thinking maybe a marine bird would snack on it.

Humans kill about 100 million sharks every year

Globally, there are fewer sharks than ever. Humans kill an estimated 100 million sharks per year. As sharks have been considered a threat to humans, they are often simply shot at or otherwise killed for no reason other than what ignorant people call "fun" or "sport."

"For sharks, the odds aren't good. The International Fund for Animal Welfare reports that every year humans kill around 100 million sharks. It's a devastating amount, especially considering that sharks are incredibly important for the overall health of the oceans.

According to one 2015 study, more than half of all Americans suffer from galeophobia. It sounds a little like a rare disease, but it's actually not a physical ailment. It simply means to be terrified of sharks."
– Charlotte Pointing, *How Many Sharks are Killed a Year*; VegNews.com; June 2023

"Humans kill an average of almost 274,000 sharks every day, over 11,000 sharks every hour, and around three sharks every second.

In contrast to the staggering number of human-induced shark deaths, fewer than 10 people worldwide are killed by shark attacks. For context, every year approximately 24 people die after being hit by flying champagne corks; toasters kill an estimated 700 people, and lightning strikes kill around 2,000 people.

More than one-third of shark species are threatened with extinction. Populations of sharks in the open ocean have declined by 71% over the past 50 years.

Many sharks are killed before they have a chance to reproduce, making it difficult for their populations to rebound."
— International Fund for Animal Welfare; April 2022

How humans kill sharks

One of the most common ways sharks are killed is by removing their fins. This usually involves cutting off the fins and letting the sharks then die as they sink in the water, unable to swim.

Bycatch

Another way sharks die is as "bycatch." This is a term used to describe species accidentally caught when fishing for other commercially-desirable marine life. Sometimes, more sharks and other creatures are caught than what the fishers are hunting.

Bycatch not only includes sharks, but rays, dolphins, turtles, seals, and other sea life – which then die, or are killed (sometimes stabbed, shot, hit, crushed, left in nets, not put back in the water, or they die in some other way). There is also freshwater bycatch, which involves another spectrum of marine life.

Some bycatch ends up being sold to the farmed animal feed industry, which turns it into "high protein" feed for cattle, pigs, and other farmed animals. Of course, cows and pigs would never naturally eat sharks, rays, and other sea animals.

One-third of shark species on verge of extinction

One-third of the hundreds of shark species are on the verge of becoming extinct.

The Gulf of Mexico oil gusher caused by the 2010 Deepwater Horizon oil disaster resulted in the deaths of unknown numbers of sharks. Years later, sharks continue to die from the poisons and other matters associated with the disaster.

Sharks play a role in our oxygen: every breath you take

Uninformed people think it is good to have fewer sharks. Do those people like to breathe?

Sharks are predators that hunt other sea creatures. The actions of sharks move huge schools of fish around in the oceans, which helps to nourish the oceans through both defecation and the decomposition of fish carcasses. The nourishment this brings to the oceans helps the kelp, sea grasses, and other marine plants grow. Marine plants provide most of the oxygen on the planet. It is the oxygen that you, me, and the rest of humanity and other wildlife depend on for life. In that way, humans are reliant on healthy populations of sharks. Even humans who live far from the oceans are breathing oxygen produced because of shark activity.

Every breath you take partially has something to do with marine life, including the activities of sharks. Every breath containing pollution has to do with humans. Be grateful to wild animals and plants for the air you breathe. As far as the pollution you breathe, blame humans.

Negligent and ignorant abuse of the oceans

"It is immoral to needlessly damage a remote and largely unknown assemblage of organisms – even if they are out-of-sight, out-of-mind, and apparently of little importance to the general ecological processes in the ocean – through negligent and ignorant abuse of the oceans."
– Martin Angel, *Ocean Trench Conservation*, 1982

"Our bridge to the future is disappearing under a flood of ignorance.

As long as this society is unwilling to make massive modifications to its ecocidal lifestyle, some of which go beyond our wildest imagination, this civilization is doomed to enter a rapid collapse full of hunger, global conflict, and mass death."
– George Tsakraklides, biologist and author of *In The Grip of Necrocapitalism: The Making and Breaking of A Psychonomy*

City pollution flowing into the oceans

As I write this paragraph, I am sitting on the end of a pier overlooking Santa Monica Bay. It is an interesting vantage point to write about the health of the oceans. This is one of the many areas of the oceans directly affected by a city built on the edge of it. Just over a century ago this bay was a pristine oasis for a variety of wildlife. For the past several decades the bay has been the focus of environmental groups working to halt toxins from flowing into it.

The city of Santa Monica has built a $12 million urban storm drain runoff treatment facility that can treat an estimated 350,000 gallons of street runoff every day, intercepting pollution that would otherwise go directly into the ocean. Plastics and other solid waste are removed and taken to a landfill. Some of the treated water goes to the local police department's plumbing, where it flushes the toilets. Other treated water is used to irrigate the city parks and cemetery.

Despite the efforts to protect the Santa Monica Bay, the water quality here often gets low grades. While other coastal cities are considering ways to treat urban-runoff, as of 2023, the Santa Monica Urban Runoff Recycling Facility remains the only urban runoff treatment plant in the country. During heavy rains the facility backs up and the polluted water is released directly into the ocean. Because of high bacterial levels, the beaches along this bay often get posted with signs warning swimmers and surfers the water is unsafe.

It all goes into Santa Monica Bay

Santa Monica Bay has been used as a dump for the military and various companies looking to unload various junk. When it rains in Los Angeles County, the trash and grime from the thousands of miles of the region's streets flow into storm drains that empty into the ocean, contaminating everything living in the water. Santa Monica Bay is often ranked as the dirtiest ocean area of western North America.

Ocean animals killed by nets, ropes, plastic, trash, and toxins

"The U.N. Environment Program estimates that 46,000 pieces of plastic litter are floating on every square mile of the oceans...

An estimated one million seabirds choke or get tangled in plastic nets or other debris every year. About 100,000 seals, sea lions, whales, dolphins, other marine mammals, and sea turtles suffer the same fate."
– Usha Lee McFarland, *Altered Oceans: A Chemical Imbalance*; *Los Angeles Times*; Aug. 2006

As I look at the murky water below the Santa Monica Pier, I can see many bits of trash floating in it. I wonder what the people who are fishing off the end of the pier think of all the pollution here. If I were someone who ate sea creatures, I would lose my appetite looking at this water. Unfortunately, the pollution here does not stay in one area.

Oceans are all connected, and share pollution

Because the oceans of the planet are really one big body of water with merging currents, what we toss into this ocean in the form of toxic chemicals and garbage often shows up thousands of miles away in another part of another ocean. As fish migrate to various parts of the oceans, just as birds migrate across continents, the pollutants fish pick up in one part of the ocean is spread to another part of the ocean, and into the other forms of life that feast on the migrating fish.

Brown pelicans and DDT

As I look across the water I see a brown pelican skimming the surface. This is a very good thing to see.

123

By 1992 the brown pelican species living off the coast of Los Angeles County was one generation away from extinction. The DDT pesticide – which is now outlawed in the U.S. – caused the birds to lay eggs with shells so thin they would break in the nest.

By 2006 it was estimated there were 7,000 breeding pairs of brown pelicans in California. By 2009, the bird had been removed from the endangered species list.

By 2022, there were an estimated 70,000 nesting pairs of brown pelicans along the California coast. Finally, a success story.

The white croaker caught in Santa Monica Bay are still unsafe to eat because of the DDT in their systems, and this is still happening over twenty years after DDT was outlawed.

Unfortunately the pollution issues of Santa Monica Bay are nothing compared to what other regions of the world's oceans are facing.

Mercury and heavy metal pollution in oceans

In March 2004 the U.S. Department of Health and Human Services as well as the U.S. Environmental Protection Agency released information detailing the levels of mercury and other heavy metals in the tissues of fish commonly eaten by humans, such as tuna, shark, swordfish, king mackerel, and tilefish, as well as shellfish. Mercury occurs naturally in the environment, but the increase in mercury in the tissues of waterlife is largely the result of industrial pollution. Much of it is pollution from coal-burning power plants, and from concrete processing kilns. While the press release mentioned how pregnant women, nursing mothers, and small children should avoid some types of fish, it stupidly also said people should continue eating fish and shellfish – likely a way to appease the multibillion dollar fishing, seafood, supermarket, and restaurant industries. Yes, government policy is influenced by corporate profits, and corporate donations to politicians.

"Unfortunately, what fish does contain is enough mercury to help you take your temperature."
– Former Montana cattle rancher Howard Lyman, in his book *No More Bull: The Mad Cowboy Targets America's Worst Enemy: Our Diet*

The high levels of mercury and other heavy metals in fish are among of the many reasons why it is unhealthful to eat fish.

Marine creatures are exposed to all sorts of toxins that are the result of poisoned rivers emptying into the oceans; of pollution directly flowing into the oceans from coastal cities, military operations, and industries; from pharmaceuticals taken by people, and used on domesticated and farmed animals; from chemical fertilizers and pest controls spread about golf courses, lawns, farms, schools, prisons, military bases, and corporate campuses; from the cruise ship, shipping, and fossil fuels industries; and from air pollution.

Bioaccumulation of toxins in marine life

When larger forms of sea creatures eat the smaller sea creatures, the toxins accumulate in the tissues of the larger fish and in sea mammals – and the sea birds and land mammals feeding off the sea life.

Pollution-tainted seafood and human illness

When a person consumes tuna, shark, red snapper, and other large fish, they are exposing themselves to all of the toxins accumulated in the tissues of those larger fish. As the U.S. government has reported, this exposure can lead to nerve damage, miscarriages, learning disabilities, birth defects, and cancers in humans.

Pregnancy and eating seafood

Mercury concentrates more in the blood of fetuses than in pregnant mothers; this is why those who are planning on becoming pregnant and those who are pregnant are advised to avoid eating tuna, king mackerel, shark, swordfish, and tilefish. Babies whose mothers consume fish are at greater risk. A plant-based diet is lower risk.

According to the Centers for Disease Control and Prevention statistics released in 1995, based on blood surveys of women of childbearing age, about 5.7 percent of infants in the U.S. could be at risk of mercury poisoning absorbed from their mothers during pregnancy.

"Fetuses, infants, and young children are most at risk from mercury exposure, because small amounts of mercury can harm the developing brain and nervous system."
– Minnesota Department of Health

"High-level exposures to mercury can cause severe birth defects, including blindness, deafness, cerebral palsy, and mental retardation, and can also result in death. In low doses, mercury may delay a child's walking and talking, shorten attention span, and cause learning disabilities."
– Missouri Department of Natural Resources, *Human Health Effects of Mercury*; MO.gov

Solid white albacore tuna has been found to be especially high in mercury contamination. One single tuna sandwich may contain enough mercury to interfere with a child's learning, concentration, behavior, coordination, and language. Because their bodies are smaller and developing, children should abstain from eating these fish.

Fish today absorb more – and a wider variety of – pollutants. These include heavy metals, microplastics, pharmaceuticals, and industrial chemicals, including carcinogens, and endocrine disruptors.

Government manipulation by seafood industry

The seafood industry continually pressures the FDA to refrain from taking actions that may interfere with profits. For instance, in meetings held between the FDA and executives in 2000, the executives expressed concerns that the FDA advisories against the consumption of tuna may result in class action lawsuits from consumers.

[An advisory released by government agencies telling people not to eat certain types of seafood] "could have an irreversible impact on American dietary habits, profoundly affecting consumers and producers of seafood and resulting in significant segments of the population turning away from the proven health benefits of fish consumption."
– The National Food Processors Association, the National Fisheries Institute, and the U.S. Tuna Foundation, in a letter to FDA, 2000; quoted in *Balancing Interests, Agencies Issue Guidance at Odds with EPA Risk Assessment: A Schoolboy's Sudden Setback*, by Peter Waldman, The *Wall Street Journal*; Aug. 2005

Fish advisory altered by seafood industry

In 2001 the FDA released a revised mercury advisory that didn't mention tuna, but did mention king mackerel, shark, swordfish, and tilefish. Americans are more likely to eat tuna, and to accumulate mercury from it, than they are from the fish mentioned in the mercury advisory. The adjustment in the wording was to satisfy the industry.

"In order to keep the market share at a reasonable level, we felt like we had to keep light tuna in the low-mercury group."
– Clark Carrington of the FDA in an FDA Food Advisory Committee meeting with officials from the Environmental Protection Agency, 2003; as quoted in *Balancing Interests, Agencies Issue Guidance at Odds with EPA Risk Assessment: A Schoolboy's Sudden Setback*, by Peter Waldman; The *Wall Street Journal*, Aug. 2005

When the FDA takes actions to protect the profits of the fish industry, one has to wonder how reliable the information is released by government agencies in relation to the safety of eating fish, or any food – or substance being used on or in food, such as farming chemicals, and food preservatives, texturing agents, dyes, flavors, and scents.

The tuna industry has used advertising with cartoonish characters and jingles aimed at children. You know the song, and the imagery.

It is wise to consider the scientific findings on mercury contamination of both fresh and saltwater fish, and the health problems related to this and other toxins, and to act accordingly. Avoid eating fish – especially when pregnant, or planning on becoming pregnant – and don't serve fish to children.

Broad bioaccumulation of pollution

Fish – and all animals – are filters. Their tissues bioaccumulate and concentrate the pollution from food, air, water, cleaning products, skin and hair products, and surrounding surfaces.

Humans are continually breathing air, drinking water, and eating food, and they accumulate the substances in the air, water, and food.

The liver accumulates the toxins taken into the body, including through food, air, and fluids, and also what is absorbed through the skin. The kidneys help filter these toxins.

Fish oil isn't a health food

Those who believe there are health benefits to be had by consuming fish liver oil might want to reconsider. The oil from the body tissues of fish contains toxins the creature was exposed to. Cod liver oil is also so rich in vitamin A that regular consumption of it can contribute to osteoporosis.

Not that you need to add oil to your foods. Every cell of every fruit, vegetable, nut, seed, and seaweed contains oil. You get essential fatty acids by eating plants. That is, just as fish get essential fatty acids by eating plants, or by eating animals that eat plants.

Food oils

If you do choose to add oil to your foods, safer oils are those extracted from raw and organically grown olives, flax seeds, grape seeds, pumpkin seeds, walnuts, and raw hempseeds.

I choose to not use oils in my foods as I eat plenty of raw fruits and vegetables – including from my garden.

Oil from fish, krill, and born beings should be avoided. If you want oil in your diet, go with what is grown, not born.

Nobody needs to be eating seafood for the purpose of getting essential fatty acids.

There are no health benefits from eating seafood that can't be obtained in better quality by eating a variety of edible plant substances – especially raw fruits and vegetables, sprouts, nuts, and seaweeds. The fish get their nutrients by eating plants, or eat creatures that eat plants. There are tens of thousands of edible plants. You have plenty of choices.

The fishing industry

Pollution is only a part of the problems facing the world's oceans, and also the life existing in and around the oceans.

Longline fishing and the depletion of sea life

"Longline fishing, used to catch swordfish, tuna, and other species, may be 80 miles long and carry several thousand baited hooks at a time. Each year longlines catch and kill hundreds of thousands of other animals, including sharks and birds."
– A Voice for Animals, VoiceForAnimals.Org

"A longline is a fishing line usually made of monofilament. The length of the line generally ranges from 1.6 km (one mile) to as long as 100 km (62 miles). The line is buoyed by Styrofoam or plastic floats. Every hundred or so feet, there is a secondary line attached extending down about 5 m (16 feet). This secondary line is hooked and baited with squid, fish, or in cases we have discovered, with fresh dolphin meat. The baited hooks can be seen by albatross from the air and when they dive on the hooks, they are caught and they drown. The lines are set adrift from vessels for a period of 12 to 24 hours.

The use of longlines in international waters is not illegal in itself. However, if the lines take an endangered or threatened species [which they often do], they become illegal because the taking of an endangered species is a violation of the Convention on the Trade in Endangered Species of Flora and Fauna. International maritime law dictates that a longline that does not bear an identifying flag is in effect legally salvageable, i.e., free for the taking because it is not attached to the ship or boat that deploys it."
– Sea Shepherd Conservation Society, SeaShepherd.Org; 2006

Billions of hooks and massive nets used to kill sea life

Fishing boats place billions of hooks in the oceans every year. Additionally, diesel fuel-spewing fishing trawlers drag nets across the ocean waters, with some nets reaching to the ocean bottom. Over the years the nets have become larger and are being dragged by larger and larger boats. And there are drift nets, set out like some sort of game play to see what shows up. Some boats can carry hundreds of thousands of pounds of fish kept on ice. Many of these fishing fleets, especially from Asian and European countries, receive money from their governments in the form of subsidies to help pay for boats, fuel, and supplies. All of this goes to supply the world markets with dead sea creatures, and some living fish and crustaceans.

Fast food industry and fish depletion

While Asian countries capture an estimated two-thirds of the world's seafood supply (much of it exported to other countries), the fast-food restaurant industry sells its customers hundreds of millions of pounds of fried fish every year.

As this trend continues, fishing is taking place deeper and deeper in the seas, and resulting in more damage – not only to the types of sea life people eat, but to sea life people do not eat, and sea life other forms of sea life depend on for their existence in the circle of life.

Killing all sorts of sea life

Fishing companies constantly kill all sorts of sea life that get tangled or caught in their nets – or on their hooks, or gets hit by the boat propellers. Some fishers kill sea mammals, the sharks, and marine birds to prevent them from eating the hunted fish.

More and more of the fish being caught are smaller and younger. They often don't get the chance to grow to their full adult size because they are caught too young. This is, before they can reproduce. This practice is also playing a role in the plunge in the populations of sea life.

Fish populations are depleted the world over. Fishing fleets often ignore quotas. And fish markets and restaurants are commonly selling varieties of sea life at risk of vanishing from the oceans, including bluefin tuna.

Overfishing, also called "fishing"

"For some of the problems we face the answers are really simple. Overfishing is the result of too many fishing vessels, poor management, and the use of bad subsidies to support fisheries which otherwise just wouldn't happen, because they wouldn't be profitable. And really to deal with illegal trade in fish. We know all of the answers to those things. And to actually implement action to really, seriously change what is going on in terms of fisheries, I believe wouldn't take that much. Politically, it may be difficult. But it is certainly well within our means, and it is certainly a much simpler problem to deal with than the climate change issue."
– Dr. Alex Rogers, Scientific Director of IPSO and Professor of Conservation Biology at the Department of Zoology, University of Oxford

Trawling the oceans is murdering the marine spectrum of life

With trawlers removing all varieties of sea life from an area of the ocean, generations of sea life are killed. This includes the young and unborn of species the fishing trawlers seek, as well as a variety of species they don't want.

Some of the sea creatures unwanted by the fishing companies survive by being thrown back into the water. Many don't. Many that are caught are smaller fish on which the larger fish survive. Not only are the large fish being removed, but the food for the remaining larger fish is depleted as well.

Bottom trawling demolishes sea floor landscapes

"Bottom trawling is a method of fishing that involves dragging heavy, weighted nets across the sea floor, in an effort to catch fish. It's a favored method by commercial fishing companies, because it can catch large quantitates of product in one go.

The problem with bottom trawling as a fishing method is that it's indiscriminate in what it catches. When dragging the large, weighted nets across the sea floor, everything that happens to be in the way gets swept up in the net, too. For this reason, bottom trawling has a large bycatch impact, with many non-target species fished in the process.

This has an impact on the biodiversity of the oceans, and also means many species are being fished to the brink simply as a consequence of commercial activities, not as the target of them.

In addition to turtles, juvenile fish, and invertebrates that get swept up in trawling nets, deep-sea corals are hidden victims of trawling.

Deep-sea coral forests, thought to be some of the most biodiverse ecosystems with high degrees of endemism (species found only there), can take centuries to form. But when a trawler runs over them again and again to catch fish, they're destroyed, and so is the whole community that had formed around them.

If we want to prevent more species from going extinct, or joining the growing IUCN Red List, we must take decisive action to restrict activities known to destroy and disturb vital ecosystems that support life on Earth."

– Ellie Hooper, *What is bottom trawling, and why is it bad for the environment?*; Greenpeace.org, April 2020

Sea coral is dying

We should not be destroying sea corals, especially in such careless ways. Sea corals are endangered. More continue to die off because of ocean warming and acidification. Corals need to be protected. They are the forests of the oceans, the nurseries for a wide variety of species. Simply by their natural functions, sea corals support all other forms of ocean life, including migrating creatures, and whales.

"Bottom trawling is a globally widespread fishing practice responsible for 26 percent of the total marine fisheries catch.

Bottom trawling is a method for catching aquatic animals that involves dragging a weighted net or rigid structure from a vessel along the seafloor. It is fundamental to the supply of a multitude of food (shrimp, whitefish, flatfish) and non-food (fishmeal and fish oil) commodities. It has played an outsized role in the industrialization and globalization of the fishing sector, becoming a mainstay of fishery economies in Europe, North America, South and Southeast Asia, East Asia, and West Africa. The vast majority of the fish caught by bottom trawlers (99 percent) is caught under the jurisdiction of coastal countries, in their exclusive economic zones (EEZs).

From 14th century 'proto trawling' to modern shrimp trawling, these fisheries have been consistently associated with social conflict (particularly in displacing traditional fishing practices), environmental degradation (in terms of contact with and penetration of the seabed, as well as impacts on sensitive species), and lack of selectivity (in terms of indiscriminately catching a range of species).

Asia is the locus of fish caught by bottom trawls; 50 percent of all bottom trawled fish is caught in the Exclusive Economic Zones of Asia, or by the foreign fleets of Asian countries. China, Vietnam, Indonesia, and Morocco are the top five bottom trawling countries, as measured by average catch over the most recent decade for which there is complete data (2007-2016). China alone catches 15 percent of the total bottom trawled catch.

Distant water fishing fleets catch 22 percent of all the fish caught by bottom trawlers in EEZs. These fleets are predominantly of Asian or European origin, and fish in the EEZs of Africa and Oceania. In 34 countries – mostly in Africa – over 90 percent of the catch caught by bottom trawlers is caught by foreign-flagged vessels. These figures could even be higher, given the significant amount of distant water fishing that is thought to be illegal, unreported, or unregulated."

– A 44-page report by 40 global experts: *New perspective on an old fishing practice: Scale, context, and impacts of bottom trawling*; Fauna-Flora.org; 2021

Trawling kills whole communities of sea animals

One or two boats pull trawler nets, which may reach thousands of feet below the surface, often scraping the bottom, altering the landscape, and disrupting and killing communities of creatures there.

"Trawling destroys the natural seafloor habitat by essentially rototilling the seabed. All of the bottom-dwelling plants and animals are

affected, if not outright destroyed by tearing up root systems or animal burrows."
– United States Geological Survey, March 2016

And trawling continues demolishing sea life

"The destructive fishing practice destroys underwater ecosystems, kills marine life, and kicks up soils needed to store carbon dioxide.

When Bryce Steward dived after the toothed, steel-weighted nets of a scallop dredger rumbling over the bottom of the Irish Sea 22 years ago, he witnessed destruction he could never have seen from a boat.

'Half crabs. Smashed up sea urchins. Starfish missing some of their arms,' said Steward, a marine ecologist at the University of York. 'There was literally a trail of dead and dying things on the seabed.'

Bottom trawling – a powerful practice in which heavy nets are dragged along the floor of the ocean to catch fish and seafood – is one of the most harmful ways to feed the world. It destroys ecosystems and sweeps up unwanted marine life that gets thrown overboard.

Now scientists fear another environmental disaster bubbling under the surface: climate change.

Trawlers churning 1.3% of the sea floor stir up more carbon dioxide than the emissions of the entire aviation industry, a study published in the journal *Nature* in 2021 found. Even if only some of that makes it to the surface, the practice stops seas from absorbing as much CO_2 from the atmosphere, and prevents plant life from growing."
– Ajit Niranjan, *Trawling seabeds makes climate change worse*; DW.com, June 2022

The article above by Ajit Niranjan goes into all sorts of information relating to what trawling has been and is doing to the sea floor, marine life, and how it is connected with weather patterns, and climate change. The article quotes Joan Mayorga, a marine scientist at the University of California, Santa Barbara, saying "The sea floor is the largest carbon reservoir in the planet, and we're disturbing it."

Massive fossil fuels used to deplete marine plant and animal life

Trawling is awful in other ways. It uses up a tremendous amount of fossil fuels to run the ships, and the fuel is more polluting then car fuel. Trawling depletes marine plant and animal life. It is ruinous from the bottom of the seas, to the populations of wildlife, to the atmosphere of the planet – including the oxygen level of the air we breathe.

"Fisheries consume about 40 billion liters of fuel annually, generating 179 million tons of CO_2-eq GHG emissions (about 4% of global food production emissions). Of the major gear types used in global fisheries, bottom trawling has the highest emissions from fuel use. Seafood is often credited for being a 'more sustainable' dietary choice with regards to climate change because the average GHG emissions per gram of protein consumed are less than 1/10 those of beef. However, GHG emissions vary significantly by gear type, and fish caught by bottom trawling can rank among the most GHG-intensive foods due to the fuel use requirements of dragging a heavy net across the seafloor.

A 2017 study showed bottom-trawl fisheries emit almost three times more greenhouse gasses than non-trawling fisheries."

– A 44-page report by 40 global experts: *New perspective on an old fishing practice: Scale, context, and impacts of bottom trawling*; Fauna-Flora.org; 2021

Trawling pollution equivalent to aviation industry

The study published on Fauna-Flora goes on to say so many things, including that pollution caused by the trawling industry is "roughly equivalent to the entire global aviation industry." In other words, trawling is massively polluting.

Unlike, the aviation industry, trawling causes tremendous destruction to the oceans, and marine wildlife deaths across the spectrum of ocean life – reducing the sequestering of greenhouse gasses in sea life, and reducing the oxygen production of the oceans through marine plants.

Read any ten articles about ocean trawling, and get a better idea of how important it is for the practice to stop. First, at least nets should not touch the ocean floor. Second, all trawling and fishing needs to end in sensitive and protected zones. Third, no trawling should be done during spawning season. Fourth, trawling and other high-intensity "harvesting" of wildlife from the seas should not be done in areas where at-risk and endangered animals – including whales – are feeding, or about to feed.

The fewer species in the seas, the less healthy the oceans become, including the marine plants, and the less oxygen there is around the globe. As fewer plants absorb less greenhouse gasses, the warmer the planet becomes – with more intense and violent storms.

The benthic regions of the deep ocean

The World Conservation Union estimates that between 500,000 to 100 million species inhabit the bottoms of the seas, known as the benthic regions. Many of these species have not been charted (identified).

The rainforests of the oceans

"The deep ocean is increasingly recognized as a major global reservoir of the Earth's biodiversity, comparable to the biodiversity associated with tropical rainforests and shallow-water coral reefs.

Though only a small fraction of the oceans' ecosystems found at depths below 200 meters have been studied, research has revealed remarkably high levels of biodiversity and endemism [animals existing only on one small area]. Estimates of the numbers of species inhabiting the deep ocean range between 500,000 and 100 million.

The development of new fishing technologies and markets for deep-sea fish products have enabled fishing vessels to begin exploiting these diverse but poorly understood deep-sea ecosystems. By far the most widespread activity affecting the biodiversity of these areas on the high seas is bottom trawl fishing.

A number of surveys have shown bottom trawl fishing to be highly destructive to the biodiversity associated with seamounts and deep-sea coral ecosystems, and concluded that it [bottom trawl fishing] is likely to pose significant risks to this biodiversity, including the risk of species extinction.

Deep-sea coral and seamount ecosystems are widespread throughout the world's oceans.

Bottom trawl fishing poses a major threat to the biodiversity of vulnerable deep-sea habitats and ecosystems. Losses of up to 95-98% of the coral cover of seamounts as a result of deep-sea bottom trawl fishing have been documented.

Given the localized species distribution and high degree of endemism associated with seamount ecosystems, bottom trawl fishing is likely to pose a serious threat to a large percentage of species inhabiting these ecosystems, including the threat of extinction.

High seas bottom trawl fishing has often led to the serial or sequential depletion of targeted deep-sea fish stocks.

Approximately 80% of the high seas catch of bottom species [groundfish, prawns, etc.] is taken by bottom trawl fishing vessels.

There has been no systematic study of the geographic extent of bottom trawl fishing in relation to vulnerable deep-sea ecosystems, or the extinct of its impact on these ecosystems."

– World Conservation Union, Natural Resources Defense Council, World Wildlife Fund, and Conservation International; June 2004

Governments fighting to continue trawling destruction

In November 2006, when United Nations negotiators tried getting a measure approving strict regulations on high seas bottom trawling, Iceland led other nations to oppose the measure. Companies in Denmark, Estonia, Iceland, Japan, Latvia, Lithuania, New Zealand, Norway, Portugal, Russia, and Spain own fleets of bottom trawling boats. Countries that supported the measure to strictly limit bottom trawling fleets included Brazil, Britain, Canada, Chile, Germany, India, New Zealand, Norway, Palau, South Africa, and the U.S.

It's like bulldozing the ocean floor

"When the trawler drags the net, it acts like an ocean bulldozer that wipes out everything in its way, from seagrass to fish, prawns, sponges, corals, algae, etc.

Very often, spawning and nursery grounds of many fish inhabit the continental shelf, which is where the trawlers tend to operate.

Bottom trawling is currently not restricted enough and serves as one of the main sources of catches. Since bottom trawls do not distinguish between target fish and non-target fish – in other words: fishers looking for a certain type of fish – the result is a high rate of unwanted fish hauled up, only to be thrown back to the sea dead or dying [or used as bait, or sold to companies that make livestock feed]."
– Oceana

Trawled creatures sold into the international market

Even where trawling has been banned, the laws are being ignored – so that companies can sell fish to stores, restaurants, catering companies, hotels, resorts, country clubs, airlines, and cruise ships. And to sell sea creatures to be made into dog and cat food, and feed for farmed animals.

It's disgusting that people refuse to obey the laws meant to protect rare and endangered sea life, and the waters and landscapes they exist in and around. It's not a small problem, it's an enormous one. It's destroying the oceans, and it plays a role in climate change and a warming planet with fewer forms of life, bringing others closer to extinction – including humans.

Trawling in the Mediterranean

"Greater transparency needed to stop illegal fishing in the Mediterranean.

Members of the Med Sea Alliance, a diverse coalition of nonprofit organizations, today launched a new data atlas which, for the first time, maps areas permanently closed to bottom trawling across the Mediterranean, and investigates trawling in these areas.

The Atlas is an online tool that maps presumed and confirmed infringements of bottom trawling in areas where it is permanently banned to protect sensitive habitats and depleted fish stocks. The Atlas has been released ahead of the 45th meeting of the General Fisheries Commission for the Mediterranean, the fisheries management body responsible for the Mediterranean.

135

In the period January 2020-December 2021, the Atlas recorded incidents of possible bottom trawling in 35 closed areas by 305 different apparent vessels across 9,518 apparent days of fishing activity (based on Global Fishing Watch data), and 169 cases of confirmed infractions between 2018 and 2020, based on MedReAct research on media outlets and information released by national control authorities.

Eighty confirmed infringements were found in the GFCM Fisheries Restricted Areas during the analyzed period. Illegal fishing clearly happened in those two years, and authorities acted and sanctioned the vessels involved.

The evidence of potential and confirmed cases of bottom trawling in closed areas suggests illegal, unreported, and unregulated fishing is undermining its sustainability, at a time when other stressors like overfishing, climate change, and pollution are already taking a toll on fish populations."
– Dave Poorvliet, *New Data Reveals Bottom Trawling In Protected Areas*; Global Fishing Watch; GlobalFishingWatch.org; Nov. 2022

Damages marine conservation efforts

"Illegal fishing in protected areas undermines national and regional management measures of fish stock, threatens the livelihoods of fishers who follow the rules, and damages marine conservation efforts."
– Aniol Esteban, Med Sea Alliance Steering Committee; Nov. 2022

Reduced genetic pool of wildlife

With at least hundreds of cases of trawling infringements on protected areas, it is clear that millions of fish are being removed there. Not only does it damage marine conservation efforts, it reduces the genetic pool of species, and eliminates some. It reduces both the chances of wildlife recovery, and the food for wild animals.

The worst fishing technique in the world

"A growing number of voices are calling for an end to what has been described as 'the worse fishing technique in the world,' bottom trawling. They are also calling for a profound transformation of this sector, even though around a quarter of the fish consumed in the world are caught using this method, which causes serious damage to the seabed.

To provide a solid foundation for regulating the practice, some 40 scientists, NGOs, academics, and environmental consultants joined forces to produce a report, *New Perspectives on an Old Fishing Practice: Scale, Context, and Impacts of Bottom Trawling*, published December 2021.

Trawling is often carried out less than 12 miles from the shore, where boats make 20 percent of their catches, putting the industry's heavyweights in direct competition with small-scale fishers whose survival often depends on the resources caught at sea. This is particularly evident off the coast of Africa, which is the number one victim of bottom trawling: over 90 percent of legal catches are made in the Exclusive Economic Zones of 34 countries by foreign vessels, particularly those from China or Vietnam, but also from countries such as Morocco, the United States, or Argentina.

These fishing giants wreak havoc on the areas where they operate."
– Annick Berger, *Will 2022 mark a turning point in the regulation or banning of bottom trawling?*; EqualTimes.org; March 2022

Trawling kill machines continue 24/7 in the oceans

In 2023, few areas of Earth's seas have limits placed on bottom trawl fishing. The destruction of marine life and the necessary terrain for its survival continues all day – every day – in the global seas.

Not only does trawling alter the sea floor, damage reefs, kill a wide variety of species the fishing teams are not aiming to catch, as mentioned, trawling boats are highly-polluting. Many travel hundreds or thousands of miles, including to go to other continental shores, where they capture huge amounts of sea creatures to sell into the international market. Often, they trawl in areas where local fishers depend on making their livelihood to sell at or to their local markets.

Trawling helps spur the crimes of piracy

"If you look at Somalia, where industrial fishing has fished out Somalian waters and the local fishermen can't get food anymore, what do they do? They turn to piracy. Who does that affect? That affects anybody with a ship that's going through those waters. They've taken a lot of different ships hostage. So, ultimately, food insecurity can become a national security issue."
– Jackie Savitz, senior scientist and chief strategist for the international ocean conservation and advocacy organization Oceana, na.oceana.org; July 2011

Trawling is largely paid for by government subsidies

Many countries subsidize the fishing industry with tax exemption for fuel, and subsidies to pay for the boats and equipment. The result is enormous fishing boats traveling far to capture massive numbers of sea creatures, and often off the shores of other countries. Local fishers, environmental groups, and the United Nations have been calling for a ban on subsidies for the trawling industry. But there are large corporations paying off politicians, and so forth. The fishing industry influence runs deep, including because some politicians – or their families – make money from it. In 2022, the U.N. estimated the international subsidies going to trawling to be $35 billion.

"The issues is 'complicated and thorny,' according to the U.N. agencies. 'For the majority of fisheries subsidies, there is a strong correlation with overcapacity and overfishing.'
 – *At Ocean Conference, U.N. agencies commit to cutting harmful fishing subsidies;* United Nations Department of Economic and Social Affairs; 2022

The correlation between trawling and the destruction of Nature is about money, profit, and greed. It is driving species closer to regional or complete extinction, and is playing a role in climate change.

Trawling needs to end. It's a rather new practice, not needed for human survival. It caters to restaurants, stores, resorts, hotels, cruise ships, catering companies, and to aquaculture fish farms, largely so continental people – who have many other food choices – can eat fish.

The trawling continues largely without limits, and remains a tragically destructive practice in the seas, where fishing companies do not obey – or skirt – laws, and every season is open season.

Micro marine life

Furthering the problem in the oceans is that food for the smaller fish is in trouble.

Phytoplankton

Phytoplankton is the food for a variety of sea life. One of the main sources of food for larger fish is the foot-long menhaden, and the chief source of food for the menhaden fish is phytoplankton.

Overfishing of menhaden

When there is too much phytoplankton, which occurs when menhaden populations are overfished, as they currently are, the phytoplankton block sunlight from reaching aquatic plants that support a variety of other sea creatures, such as oysters that help filter the water.

Urban and farm pollution spurs algae dead zones

Because water flowing into the oceans is loaded with landscape chemicals, farming chemicals, and the urine and feces of farmed animals, the coastal waters are out of balance with algae overgrowth. Much of this is the result of fertilizers used on lawns, sporting fields, and massive fields of feed crops grown to feed billions of farmed animals.

Water oxygen depletion and dead zones

As the populations of phytoplankton-eating sea life are on the decrease, and algae are on the increase, the coastal waters, and increasingly large areas of water away from the coasts, are being devastated by algae overgrowth. This robs the water of oxygen, causing massive fish kills. The rotting vegetation sinks to the bottom where it kills even more sea life, including because of the bacteria feeding on it absorbs oxygen, resulting in dead zones.

Oysters dying off

The dead and dying at the bottom of the seas include oysters, which have already been overfished, killed off by pollution, and are regionally extinct in areas where they flourished just a few decades ago.

Toxic sea life

With the oceans absorbing more and more pollution, fish are becoming toxic to themselves, and to marine and land creatures that eat them, and to humans who eat fish and shelled marine creatures.

Some nursing sea mammals have such high levels toxins in their bodies their young are poisoned from toxins concentrated in milk. Other sea mammals can't reproduce because of a lack of food.

It isn't only anchovies, bass, cod, grouper, haddock, halibut, herring, mackerel, marlin, orange roughy, pollack, rockfish, salmon, sardines, snapper, squid, swordfish, tilefish, tuna, whiting, and other commonly eaten sea life being killed by fishing and found tainted by pollutants.

Whales, dolphin, porpoises, and sea turtles

Whales, dolphin, porpoises, and sea turtles are still being killed and sold for food in many regions. Some end up as food for pets, and for farmed animals.

Whales continue to be hunted by commercial fishers from Japan, Norway, Iceland, and by indigenous people in Denmark, Korea, Russia, and the U.S.

Whales slaughtered in Faroe Islands

There are also the massive pilot whale slaughters done by the people of the Faroe Islands. They kill many, many pilot whales every year. They also kill hundreds of dolphin. The murders take place when boats drive the animals into a bay, then people go into the shallow water and slit and or stab them with knives. The bay water turns to the color of blood. The people say the slaughter is "tradition."

Some traditions need to end. Especially when they are so blatantly needlessly harming wildlife.

The people of the Faroe Islands have other sources of food, and do not need to be killing whales. (Search: Whale slaughter Faroe Islands.)

"Every year in the Danish protectorate of the Faroe Islands an archipelago just 230 miles north-west of mainland Scotland, entire families of small cetaceans – primarily long-finned pilot whales and Atlantic white-sided dolphins – are massacres each year in drive hunts called 'grindadrap.' When a pod is spotted, more boats and jet skis are used to chase the cetaceans, sometimes for hours, 'driving' them into one of the many shallow bays of the Faroe Islands, where locals on the beach used hooks to pull them ashore, and kill them with spinal lances and knives.

The 'grind,' as the hunts are commonly called, can happen at any time at any one of the 26 designated killing bays around the islands, with the majority of the hunts statistically occurring between July and September, and an average of 1,156 pilot whales and dolphins killed each and every year over the past 40-year period. The grindadrap has no season, no quota, and no restrictions on killing pregnant females or juveniles. Entire families are killed in the name of tradition – a tradition that has no place in a country where today the standard of living equals its Scandinavian neighbors, and even the Faroese doctors say the meat is too toxic for regular human consumption."
  – *Operation Blood Fjords: Defending pilot whales and other dolphins in the Faroe Islands: A cruel and outdated slaughter in the name of 'tradition'*; SeaShepherdGlobal.org,

Boats surround the whales and force them toward shore

"It is unbelievably brutal and inhumane to slaughter such intelligent animals like this. Boats surround the animals and the hunters bang long metal poles under-water to create a wall of sound that disturbs their sonar and confuses them.

The boats push the dolphins into a small lagoon where they are trapped by nets. They are then left for a night or two – they're strong animals and the hunters want to tire them out.

They are harpooned from boats while some men jump into the bloody water with big knives to cut their throats. Sometimes they're hooked and lifted out of the water while still alive."
  – Clare Perry of the Environment Investigation Agency describing the slaughter of dolphin by Japanese fleets, quoted in *Slaughtered: They're Friendly, Intelligent and Our Kid's Dream of Swimming with Them. So Why Are Thousands of Dolphins Still Being Slaughtered?*, by Gary Anderson, *Sunday Mirror*, Sept. 2006 (SundayMirror.Co.UK). The article mentioned it was expected the Japanese fishermen would kill about

20,000 dolphins over the following sixth months. (See: EIA-International.Org; and MarineConnection.Org)

Japan killing sea mammals

Japan is often mentioned as the worst offender when it comes to killing sea mammals, including dolphin, and minke and fin whales, who all struggle and scream as they are mercilessly killed in ways that may take hours as the water around them turns red with their blood.

Japanese fishermen have caused great harm to populations of Dall's porpoises. In January 2002 it was reported that because the population of Dall's porpoises had plummeted, the fishermen resorted to killing porpoises that are pregnant, or are still nursing, which often results in the death of the calves through starvation or shark attack.

When the fishing boats surround the dolphin pods and use large nets to drive them into coves, the adult male dolphins surround the females and young to protect them. After the boat crews gather some of the dolphins to sell them to aquariums and tourist parks (for as much as a couple hundred thousand dollars), the remaining dolphins are stabbed in a massively bloody slaughter.

Amusement parks buying sea mammals

"The U.S. is currently not importing live dolphins, but many of the marine mammals currently on display in the U.S. were caught in the wild. We reward the industry described above when we pay for tickets to see them. When we swim with captive dolphins while on vacation in other countries we are directly supporting the slaughter."
– Karen Dawn; DawnWatch.Com; Oct. 2006

Contaminated dolphin and whale meat

The meat of one adult dolphin can be sold for about $600. It doesn't seem to matter to consumers that dolphin meat is contaminated with high levels of methyl mercury, a particularly toxic form of the toxin. Whale meat, which is also consumed in Japan, has also been found to contain unhealthful levels of the toxin.

The Taiji dolphin kill horror

"On September 1st, 2022, in Taiji, Japan, dolphin hunters once again begin the six-month-long season to chase, net, and capture dolphins for captivity. They will also slaughter the pod mates and family members of those dolphins chosen for the lucrative dolphin trade and life imprisonment. The slaughtered dolphins are 'less desirable' for captivity than those for a lifetime in small tanks.

The Taiji dolphin hunts, brought to the world's notice by the documentary *The Cove* by Louie Psihoyos and the Ocean Preservation Society, documents the efforts of the International Marine Mammal Project of Earth Island Institute to end the dolphin slaughter. *The Cove* received the Oscar for best documentary in 2010."

> – Mark J. Palmer, *Taiji Dolphin Slaughter Begins Again*; International Marine Mammal Project, SaveDolphins.eii.org

"The Japanese drive hunts are an astonishingly cruel violation of any reasonable animal welfare standards."

> – Lori Marino, Ph.D., Emory University, and Diana Reiss, Ph.D., New York Aquarium and Columbia University. These two scientists proved dolphins recognize their reflection in a mirror, a cognitive complexity elephants and chimpanzees also have.

"During drive hunts, migrating pods of dolphins and other small whales are first panicked and confused by loud banging, then herded, by the hundreds, into shallow coves and butchered, one by one, by fishermen. Every year, some 20,000 small cetaceans of several species, some of which are endangered, including bottlenose dolphins, striped dolphins, spotted dolphins, Risso's dolphins, short-finned pilot whales, white-sided dolphins, and false killer whales, are killed or taken in the drives, sometimes illegally.

This cruel and inhumane practice is sanctioned and controlled by the government of Japan, which claims that these animals compete with the fishermen and slaughtering them is a means of pest control, but no evidence for this claim exists. The dolphins are processed and used as pet food or fertilizer, and the government is encouraging the consumption of dolphin meat. In fact, the hunts would be economically unviable without the sale of live dolphins captured during the drives to dolphinariums in Asia and elsewhere."

> – TheOceanProject.Org/ActForDolphins

Intelligence of dolphins, porpoises, and whales

"Scientific research shows that dolphins are highly intelligent, self-aware and emotional animals with strong family ties and complex social lives."

> – from the Scientist Statement Against the Japanese Dolphin Drive Hunts, to the Government of Japan. The statement was signed by hundreds of scientists from around the world. TheOceanProject.Org/ActForDolphins; Nov. 2006

"You cannot ignore any longer the fact that these animals have very large brains, highly developed societies, social relationships, and sophisticated cognitive abilities."

> – Richard Connor, Ph.D., University of Massachusetts, Dartmouth

Sea net deaths of marine birds, mammals, and turtles

It is a travesty that hundreds of thousands of dolphins, porpoises, and whales are being killed by fishing teams every year; it is so much worse that many other forms of sea life are being killed for no reason other than they just happened to be in the wrong place at the wrong time. In so many areas these large sea animals, including endangered sea turtles, are being killed because they get caught in gillnets that stretch anywhere from the surface of the water to the ocean floor. The dolphins, porpoises, whales, and other sea mammals, and even diving sea birds cannot detect the nets, get tangled in them, and drown.

There are simple, low-cost modifications that can be made to gillnets to greatly reduce the number of dolphins and porpoises killed, but the fishing companies have not made the changes.

Marine life around the world is considered by many humans as if it is all there for the taking. And killing, eating, selling, and exploiting.

Norway killing minke whales

"Norway plans to kill up to 1,278 minke whales this year, according to a recent announcement made by the country's fisheries ministry. This is the same quota as the previous two years, although whalers only killed 503 common minke whales (Balaenoptera acutorostarta) in 2020, and 429 in 2019.

In 1982, the International Whaling Commission issued a global moratorium on commercial whaling, which went into effect in 1986. But Norway, despite being a member of the IWC, formally objected to this ruling, and has continued to kill whales every year."
– Elizabeth Claire Alberts, *Minke whales for dinner: Norway's controversial whale hunt is still on*; News.mongabay.com, March 2021

"Humans have a very poor track record of doing anything sustainable when it comes to whaling, and we shouldn't continue to allow those kinds of processes to occur, just because someone says they can do it sustainably. The fact that there is an international ban on commercial whaling since 1986 strongly indicates there's not a way to do it in a way that is going to be sustainable.

The more we learn about whales, the more we recognize that they provide ecosystem services at a range of scales. The presence of whales and the abundance of whales can actually stimulate growth because of how they circulate nutrients that are limiting in the environment – and if you take a lot of animals out of a small area, you can impact the productivity of that ecosystem and upset the balance of the ecosystem, too. And if you take away a significant predator that

eats a lot of the forage fish, or one certain type, it's going to leave opportunities for other species to come in and change the structure of the ecosystem, and those can have consequences we're not familiar with."
– Ari Friedlaender, University of California, Santa Cruz, 2021

Whales and sharks play major roles in nutrient distribution

Just as sharks, and other larger marine animals do, whales play an enormous role in the distribution of nutrients in the oceans, and that plays into how other forms of life in and around the oceans survive, and how much of the plant life in and around the oceans thrive, or die off, and how much oxygen they produce.

How sea nutrients get to the mountain forests

As is mentioned elsewhere, even the nutrients carried within the tissues of the salmon that swim far up into rivers helps to fertilize the forests, as the salmon are eaten by bears, wild cats, and eagles, and other carnivores who then carry those nutrients out into the land – fertilizing plant life. It's all the circle of life, the circle of plant life, the circle of oxygen, and how healthy the forests are as they and the oceans play roles in the formation of clouds that then hydrate the planet.

Whales, dolphin, porpoises, seals, sea lions, sea turtles, and other marine life need to be protected so their populations can return to healthful and sustainable levels. Including to help the forests and other plant life, the formation of soil, and the rain and oxygen production around the globe. Rivers need to be restored, and protected.

Whales are key to a functioning ecosystem

"I don't think any country should continue whaling. It's no longer a discussion about if whales are a charismatic species, or if eating whales is a cultural right, or if sustainable use of whales is correct, or not. During the last ten years, scientific information is showing us the key role of whales in the functioning of the ecosystem."
– Else Cabrera, Cetacean Conservation Center in Chile

Japan killing whales, and the government's lie about doing so

In June 2005, Japan announced to the International Whaling Commission in Ulsan, South Korea, it would increase the number of mink whales it would kill from 440 to 935. In 2018, Japan killed 640 whales. Japan claims it kills whales to study them and to "monitor stocks," but ends up selling the meat for food. The commission had banned commercial whaling in 1986.

"Norway killed 580 whales in 2022, and has reportedly hunted 15,000 whales since the 1986 moratorium was established.

The numbers of whales hunted by Japan has been decreasing most years, although the nation intends to ramp up its whaling operations. In 2018, Japan killed 640 whales, compared to 383 in 2021, and 270 in 2022. Iceland's 2019-2023 quota permitted the killing of 217 minke whales, and 209 fin whales."
– Jennifer Mishler, *Whale Hunting Still Exists in 3 Countries, but It Is Declining*; SentientMedia.org, Sept. 2023

Smaller fish being caught: before they reproduce

"According to the Marine Fish Conservation Network, north Atlantic swordfish caught today are only a third the size caught in the 1960s when I was out spotting for snappers – and well below that [weight] which females must be to reproduce.

Sea Web reports that of the 157 fish species tracked in U.S. waters, 36 percent are overexploited and 44 percent are fished to the max. Populations of cod, haddock, halibut, red drum, and yellowtail flounder are at record lows. Chilean sea bass is so overfished that many scientists predict commercial extinction within five years."
– Doug Moss, *Will the Real "Slob" Fishermen Please Stand Up?*; E magazine; July/August 2005

Dangerously small gene pool of sea life

The big fish are being killed off by overfishing to such an extent populations are less than 10% of what existed just 60 years ago. This is a dangerously small gene pool. At the same time, the small fish are killed because of net fishing; dynamite fishing; damage to mangrove forests, coral reefs, and seagrass terrain; and from plastic, industrial, city, landscape, and petroleum and animal farm industry pollution. Creatures of the lower depths are killed by overfishing, pollution, and the overgrowth of algae and bacteria. Some water plants are overgrowing because of landscape chemical and farming pollution, and this causes the death of some fish and bottom feeders. Oysters, menhaden, and other sea life that work as natural water filters are overfished, are dying because of pollution, and are suffocating because of those radical algal blooms and bacteria feeding on the dying algae.

Nearly every form of sea life is suffering

In other words, as I mention in so many ways throughout this book, nearly every form of sea life is suffering. Even those that aren't will lack access to nutrients as other ocean life is reduced, the light spectrum is changed within the oceans, plastic pollution proliferates, ocean acidity increases, ocean temperature changes, and the ability of the wide variety of marine life to survive is reduced. All of this is because of human activity.

Farmed animals fed fish, and fish used as fertilizer for feed crops

Some may argue that the non-desirable sea creatures killed by fishing companies are not a total waste as much of the kill ends up being sold to companies that turn it into feed for farmed animals, and feed for fish farms (such as shrimp farms on the rims of the Indian Ocean that have caused vast amounts of damage to the mangrove forests, and helped contribute to the tragedy of the tsunami that devastated the region in December of 2004). In addition, much of the sea life caught that can be fed to humans *also* ends up being turned into feed for farmed mammals, and for farmed fish, and also into fertilizers (mostly used to grow feed for farmed animals). Some is turned into dog and cat food.

Killing billions of animals for human consumption

Have you noticed a theme? Farmed animals. Hunted animals. Killing billions of animals for human consumption, and for farmed animal consumption, and growing massive fields of feed crops, is tilting animal and plant life toward extinction. Including the human animal.

Farmed salmon

"It takes three to five kilograms of other fish, such as herring and anchovy, to make the feed necessary to produce one kilogram of farmed salmon resulting in a loss of edible animal protein worldwide.

In Canada it is illegal to make animal feed out of proteins otherwise suitable for human consumption. As a result, most of the feed for British Columbian (farmed) salmon is obtained from South America. This reduces the amount of food energy available to people there."
– FarmedAndDangerous.Org

Damaging everything from ocean to mountain life

Today's massive fishing operations are devastating to the ecosystems of the oceans, seas, forests, meadows, streams, rivers, lakes, lakes, ponds, swamps, marshes, and wetlands.

Tyson Foods: the animal killing corporation

Who owns some of the largest fishing companies? Some of the largest animal farming companies, such as Tyson Foods. The Arkansas-

146

based company also kills millions of chickens every year. Tyson Foods is one of the most environmentally destructive companies.

Shells and oysters

Many types of sea life are not killed meat. Oysters are killed for pearls. Many sea animals are killed for their pretty shells.

Corals killed for surgery, supplements, and decorations

Some types of coral are killed for use in calcium supplements, and others are used as filler in human bone surgery. Many types of coral are killed to supply pet stores with aquarium decorations.

Silky white-tip sharks vanished from Caribbean

As mentioned earlier, many sharks are killed for their fins, which are used in Chinese medicine, and in shark fin soup. Very often, sharks are hauled onto fishing boats, their fins are sliced off, and the sharks are thrown back into the ocean where they struggle to swim until they die. Whale sharks, which are considered to be harmless to humans, are also being killed for their fins. Hammerhead and great white shark populations have been reduced by over three-fourths in the recent decades. The Caribbean is basically empty of the silky white-tip sharks once common there.

In addition to the fishing industry killing untold millions of wild fish, there is another threat to sea life.

Threat of fish farms

No story about the health of the oceans would be complete without mentioning the growing threat of fish farms.

"In 2013, the number of farmed fish exploded past capture fisheries, and only continues to climb.

Fish farming is the industrial practice of aquaculture where huge quantities of fish are bred and raised in enclosed, unnatural conditions, and to be slaughtered in a commercial setting and sold as food. Approximately half of all fish eaten around the world come from industrial farms.

Farmed fish are typically grown in large tanks, industrial enclosures, sea cages, net pens, or small ponds to produce high yields of fish in short periods of time.

It can take up to five pounds of smaller wild fish from the ocean that are fed to the farmed fish to produce just one pound of fish meat from salmon or bass, two of the most common fish being raised on

factory farms. The methods used to capture these wild fish that become farmed fish food are appalling. Trawling is a dangerous form of industrial fishing that literally scoops up anything in the net's path."
   – Grant Lingel, *Fish Farming: Harming Oceans While Poisoning People and the Environment*, SentientMedia.org; April 2019

"Salmon farming is wreaking ruin on marine ecosystems, through pollution, parasites, and high fish mortality rates which are causing billions of pounds a year in damage, a new assessment of the global salmon farming industry has found.

Taken together, these costs amounted to about $50 billion globally from 2013 to 2019.

Fish mortality has more than quadrupled, from 3% in 2002 to about 13.5% in 2019, in Scottish salmon farms alone. About a fifth of these deaths are recorded as being due to sea lice infestations, but about two thirds are unaccounted for, so the real mortality owning to sea lice – which feed on salmon skin and mucus, effectively eating the fish alive – could be much higher.

Scotland is one of the biggest producers of farmed salmon in the world."
   – Fionna Harvey, Global salmon farming harming marine life and costing billions in damage; TheGuardian.com, Feb. 2021

"Shrimp farms are the primary cause of the destruction of the world's mangrove forests."
   –VoiceForAnimals.Org

When you see fish sold in stores and restaurants, now you know what went on to get it there. The fishing industry wouldn't want you to know.

Astonishingly arrogant degree of secrecy

"Though funded with public money, the process of developing open-ocean aquaculture has been conducted with an astonishingly arrogant degree of secrecy."
   – Ben Belton, *The Ecologist* magazine; July-August 2004

Pollution from farmed fish

"Imagine the raw sewage that half a million people would create in one day. It is probably too much to imagine. Now imagine if it were pumped directly into the ocean without having been treated. There are presently over 85 open net cage fish farms currently operating in the coastal waters of British Columbia producing waste that is equivalent in volume to the raw sewage released from a city of 500,000 inhabitants."
   – FarmedAndDangerous.Org

Increasingly serious environmental problem

Farmed fishing has been a quickly growing form of producing products for the international grocery store, catering, airline, hotel, resort, cruise ship, and restaurant markets. It is an increasingly serious environmental problem. It damages or destroys coastal mangrove forests, marshes, beaches, and wild fish populations, and results in the deaths of many thousands of sea mammals and waterfowl.

Farmed fish colonies create huge amounts of pollution – not only from the fish waste, but also from the chemicals used in the industry. The water is treated with herbicides to prevent water plant growth. The nets used to surround the fish farms are damaging to the environment as the nets are often treated with chemicals to prevent the growth of sea organisms – including barnacles and mussels – which encrust the nets. Other chemicals are used to prevent parasites and other situations.

Salmon and bass are the two most common forms of fish raised in industrial fish farms.

Pellets containing dyes to color the fish meat

With salmon farming, the pellets fed to farmed salmon contain dyes to make the flesh of the salmon take on the color wild salmon get from eating their natural food, krill.

Issues of concern related to fish farming

Pollution from fish farms altering wildlife

- Pollution from the fish farms altering nearby waterlife, and/or killing it. This includes the amount of fish poo produced in such a small area among thousands of fish in cages floating in the oceans. The waste from fish farms also increases algal blooms in nearby water, increasing the amount of bacteria in the water, decreasing oxygen levels, and suffocating natural marine life.

Pharmaceuticals and pesticides used on farmed fish

- Chemicals – including antibiotics, pesticides, and toxic substances – used on fish farms damage surrounding waterlife.

Toxins found in farmed fish

- Farmed fish have been found to contain higher levels of PCBs than do wild fish. This is partially because larger farmed fish are also fed fish meal made of smaller fish also raised on industrial fish farms. Other industrial chemicals include dioxins, dibutyltin, and polybrominated diphenyl ether. Also pesticides, insecticides, miticides, and heavy metals, including mercury. The toxins accumulate in stronger amounts each step of the way up the food chain. Then, humans eat those contaminated fish. The chemicals and heavy metals, or combinations of them, can cause cancer, hormonal imbalance, nerve problems, brain function issues, birth defects, autoimmune disorders, and other health problems.

Ocean ranches exploit waterlife and environment

- The governments sell off parts of their Exclusive Economic Zones (EEZs) in the oceans to companies and private investors and allow the "ocean ranches" the right to commercially exploit the water, waterlife, and minerals within those zones. This is especially common among poorer countries.

Diseases and parasites developing in fish farms

- Diseases developing in the fish farms, and spreading to nearby wild fish populations. This includes lepeophtheirus salmonis sea lice infestations. These parasites feed on the fish, essentially eating them alive. Nearby wild fish have been found with the lice. As they travel, they can spread the parasites far and wide among other wild fish.
  Farmed fish are more likely to be stressed, and have weakened immune systems, increasing their susceptibility to illnesses and disease.

GE fish escaping into wild

- Genetically modified fish in fish farms escaping into the wild and contaminating wild fish, and, as an invasive species, preying on fish of already compromised populations.

Fish farmers killing marine birds, mammals, and turtles

- Fish farmers killing the sea mammals and marine birds showing up to feed from the farmed fish. Among the animals being killed by fish farmers are gulls, eagles, cormorants, turtles, beavers, otters, seals, and sea lions,

Fish farms rely on highly ruinous trawling

- The numbers of wild fish that need to be caught to feed the farmed fish. This includes trawling the oceans with massive capture nets that kill all sorts ocean creatures not wanted, and damages the sea floor, corals, and the delicate ecosystems that support all ocean life.

Antibiotics used on fish farms spreads resistance

"Overuse of antibiotics – in farming or for human medical treatment – speeds up the development of antibiotic resistance, which is when bacteria change and become resistant to the antibiotics used to treat infections they cause. This is compromising our ability to treat infectious diseases, and undermining many advances in medicine."

– Dr. Danilo Lo Fo Wond, Program Manager for the Control of Animocrobial Resistance, World Health Organization

Contaminated fish are sold in stores and restaurants

"Seven of ten farmed salmon purchases at grocery stores in Washington DC, San Francisco, and Portland, Oregon were contaminated with polychlorinated biphenyls (PCBs) at levels that raise health concerns."

– Environmental Working Group

Various types of fish farming

Farmed fish aren't only grown in floating cages in the oceans, but are farmed in lakes, ponds, irrigation style ditches, cement enclosures, river enclosures, plastic tanks kept in greenhouse structures, tanks inside industrial warehouse type structures, and on coastal land.

Body chemistry altered by fish in a containment

Besides the parasite and chemical pollution and waste issues, each form of fish farming presents problems. One is of stressed and frustrated fish kept in the incredibly crowded enclosures. This situation alters their chemistry and fat content. What people eat alters their chemistry. What does eating sickly fish tainted with industrial chemicals do to humans?

Farmed fish toxins cause human health issues

Farmed fish are more likely to carry increased levels of toxic chemicals and heavy metals known to cause or worsen such conditions as birth defects, miscarriages, immune disorders, learning disabilities, nerve damage, and various types of cancer.

Cancer risks of eating farmed fish

"The groundbreaking study, *A Global Assessment of Organic Contaminants in Farmed vs. Wild Salmon: Geographical Differences and Health Risks*, was released January 2004 in the respected journal *Science*. The study, which is being considered the most thorough analysis of farmed and wild salmon to [that] date, found in most cases that

151

consuming more than one serving of farmed salmon per month could pose unacceptable cancer risks, according to U.S. Environmental Protection Agency (EPA) standards for determining safe fish consumption levels. Farmed salmon were found to have up to ten times higher levels of PCBs and dioxins than wild salmon."
– FarmedAndDangerous.Org

The meat of the farmed fish is less healthful than fish from the wild. Farmed fish meat is higher in saturated fat, which is one reason it is also higher in chemicals. Environmental pollutants gather in fat. The antibiotics given to the farmed fish contribute to human health problems, especially through the increased likelihood of drug-resistant bacteria.

Fish farms razing coastal forests and spreading toxins

"The dense salmon farms of Canada and northern Europe helped spread disease among wild fish while releasing waste into coastal waters. Mangrove forests, which provide a valuable habitat for coastal life, have been razed to make way for Thailand's shrimp farms. Especially troubling, many of the most popular farmed species are carnivores, meaning they need to be fed at least partly with other fish. By one count, about two pounds of wild fish ground up to make fish meal is needed on average to produce one pound of farmed fish, which leaves the ocean at a net loss.

As production pressures have ramped up, Chinese manufacturers have packed their ponds more tightly, leading to disease and pollution from fish waste. That waste can overload coastal waters with nutrients, causing dead zones that can strangle sea life. To fight the diseases worsened by crowding, Chinese fish farmers have liberally used antibiotics and other drugs, including malachite green, an antifungal agent and potential carcinogen that was banned by Beijing in 2002 but shows up periodically in exports."
– Bryan Walsh, The End of the Line; *Time Magazine*; July, 2011

Sea lice, other parasites, viruses, and bacteria

Farmed fish are more likely to become diseased through bacteria, through sea lice and other parasites, and through viruses. This includes the highly contagious virus infectious hematopoietic necrosis, which contaminates salmon farms. These diseases and infestations are serious risk factors for the wild fish swimming or feeding near the fish farms.

Damaging to local wild fish and other animals

It has become increasingly common for fish farms to be growing types of fish not native to the region of the farms. When these non-native fish escape into the wild, they compete with native fish, and in some cases kill or otherwise damage the populations of wild fish, as well as surrounding ecosystems.

Fish farmers also kill predator sea life to stop them from eating the farmed fish.

Hundreds of thousands of seals, sea lions, others killed

Fish farmers along the coast of British Columbia have reportedly killed hundreds of thousands of seals, sea lions, and otters.

Deterrents damaging to whales, dolphins, porpoises

Some fish farmers also use underwater noisemakers to deter whales and other marine mammals from entering the surrounding waters. This interrupts the migration patterns of the sea mammals. As the acidity of the oceans increases, the sound waves in the oceans travel greater distances, and are louder. This can also cause damage to whales, dolphins, and porpoises.

Using lights to keep fish eating day and night

Some fish farmers use bright lights to keep the fish eating all night so they become fatter faster. This confuses wild fish, alters their body chemistry and fat content, and contributes to the damage of the surrounding ecosystem.

Sending fish to market: pain, agony, suffering, and death

When it comes time to send the fish to market, there are a variety of ways they are killed or packaged.

Fish do feel pain, and struggle and fight to escape stressful, uncomfortable, painful, or dangerous situations.

What type of fish they are, how many there are, and how they are to be sold or shipped determines how and when they will be killed. Some are shipped alive to distant locations, processing plants, or sold to markets still alive in some form of container.

Depending on the conditions of and containers used for shipping live fish, they might be physically injured and/or deprived of food, increasing their suffering, increasing their stress chemistry, and altering their fat.

Some farmed fish simply are removed from water, suffocating them – which can take more than several minutes. Some are cut, which could mean minutes of bleeding to death as they also suffocate. Others are put into a slurry of ice, which can kill them in a variety of ways, including ice becoming lodged in their gills and mouths – and this can take ten – or more – minutes to kill them. Some are electrocuted, which might only stun them, leaving them conscious as they are cut apart and packaged for shipment. Some are clubbed or hit, a gruesome way to die.

Fishing industry wreaks havoc on marine life

The entire fishing industry has wreaked havoc on populations of all sea life – from the lobsters, shrimp, and other shellfish to sea birds and sea mammals, including turtles, seals, sea lions, dolphins, porpoises, and whales.

The populations of most types of creatures of the sea have been decimated, and some have become regionally extinct as the result of the wild capture and containment fishing industries. Unknown numbers of endemic sea life (that existed in one small region) have gone extinct because of fishing, terrain destruction, and pollution.

Overfished

Across the board you can look at all types of water creatures humans consume and see that not only are every one of them "overfished," but also their populations are continuing to decline. Often, they are being caught younger, and before they are able to reproduce.

Regional extinction of some fish because of humans

Not only are populations of sardines, mackerel, and herring being fished to regional extinction, nearly all forms of sea life have been reduced, and many have become extinct in the past 100 years. Others are endangered, or are on the edge of becoming extinct, such as beluga sturgeon, which are killed for their eggs sold as caviar.

Caviar market causing sturgeon populations to plunge

"The Beluga sturgeon have been put on the Endangered list of species as they have been relentlessly hunted for caviar, which is one of the most expensive caviars in the world due to scarcity. Beluga sturgeon caviar is banned in the United States due to its endangered status.

Although the importation of Beluga caviar is banned in some places, unrealistically high annual harvest quotas continue to threaten the species in its natural environment. Some estimates suggest the species soon could be extinct in the wild."
– AZ Animals

Sturgeon are on the Red List of Threatened Species

"Eighty-five percent of sturgeon, one of the oldest families of fishes in existence, valued around the world for their precious row [eggs], are at risk of extinction, making them the most threatened group of animals on the IUCN Red List of Threatened Species. The latest update of the Red List assessed the status of 18 species of sturgeon from all over Europe and Asia, and found that all were threatened.

Twenty-seven species of sturgeon are on the IUCN Red List with 53 percent listed as Critically Endangered, the Red List's highest category of threat. Four species are now possibly extinct."

– International Union for the Conservation of Nature, *Sturgeon more critically endangered than any other group of species*; IUCN.org; May, 2010

"Along with five other sturgeons in the Danube, Beluga sturgeon numbers have collapsed in recent decades due to overfishing and dams blocking their migratory routes. The gravest threat to their survival is poaching to supply Europe's flourishing illegal trade in wild caviar and meat.

Within the EU, the Danube is the only river remaining with naturally reproducing sturgeon populations."

– Panda.org, June 2020

"Of the 27 described species of sturgeon, 26 are now at risk of extinction (this means they are classed as Vulnerable, Endangered, Critically Endangered, or Extinct in the Wild). The 27th, the Chinese Paddlefish, is described as Extinct. The Yanatze sturgeon (Acipenser dabryanus) was reclassified from Critically Endangered to Extinct in the Wild. All the A. dabryanus currently inhabiting the Yangtze River were artificially propagated and released; no wild-born Yangtze sturgeons exist anymore. There has been no evidence of natural reproduction for this species since 2000. Given the perilous situation of the Yangtze sturgeon, the species was recently included in a joint publication by the IUCN SSC Freshwater Conservation Committee, Shoal, and the Indianapolis Zoo on 'Fantastic Freshwater' which highlights the diversity and beauty of 50 freshwater or wetland-dependent species and the threats they and their habitats face.

Sturgeons are an ancient lineage of fish. Having originated in the late Triassic period at a time when dinosaurs still roamed the Earth, they have changed very little in appearance over the course of their evolutionary history, and they are among the oldest and largest fishes in existence. Some sturgeon have reached up to 8 meters in length.

Sturgeon have been overfished for their meat and caviar for centuries. With sturgeon fishing now prohibited practically everywhere, the approximately 500 tons of caviar produced annually stem from farmed animals – yet poaching still impacts more than half of those species. Given that they are long-lived species and don't reach maturity until 10-20 years of age, these magnificent fish have long generation lengths – this means the average age of individuals producing new offspring is relatively high and it takes a longer time for the populations to replace itself."

– Global Center for Species Survival, *Ancient Fish Threatened with Extinction*; July 2022

Poaching of fish and other animals

When people hear the word "poaching" they often think of elephants who are killed for their tusks, but many animals are hunted and killed for their various parts or substances that can be sold on the underground market. This includes the eggs of sturgeon, and the fins of sharks.

Other commonly poached animals include every type of leopard, pangolins, mountain gorillas, lemurs, the Burmese star tortoise, lions, tigers, bears, eagles, and every type of rhino.

People need to stop killing wildlife. It's as simple as this: each species removed from the circle of life causes another type of plant or animal to enter the trajectory to extinction.

Ocean life is exhaustible

"We thought the ocean was an inexhaustible source of food, and that it would always regenerate more creatures to replace those we remove. But today's ocean is failing to produce fish as it did in the past, and the reason increasingly appears to be an overall decline in marine nutrient cycling. Simple starvation now increasingly limits the growth of fish and other marine life. 'Overfishing' appears to have affected not only the targeted species, but also the ecosystem in general... and ramifications of this disturbance may extend as far as the ocean-atmosphere CO_2 balance."

– *The Starving Ocean*, FisheryCrisis.Com; 2005

When waterlife is damaged, so is all life

When waterlife is damaged, all life is damaged. John Robbins explains this in his book *The Food Revolution*. As I've mentioned, fish of all sizes play a critical role in the existence of all other sea life. When one type of fish is overfished, or becomes extinct, it disrupts the ecosystem. This includes both saltwater and freshwater fish. And it impacts not only life within the water, but also the lives and population of birds, land animals, insects, and bugs. It also affects the plants and trees in the forests; this is because as, is mentioned earlier, the animals that feed on waterlife defecate on land, and this nourishes the plants, fungi, and bacteria. When the animals die, their decomposing bodies then turn into nutrients that feed the other life.

For humans to healthfully exist, they need all of the animal life behaviors to carry on, generation after generation, among wildlife thriving in their habitat.

Marine carbon and isotopes help forests

Marine carbon and nitrogen isotopes are two beneficial atomic nutrients brought into the forests by birds, bear, wildcats, foxes, and other animals that consume wild fish.

Unless humans make enormous changes: the ocean collapses

A global study authored by 14 marine biologists published in the November 3, 2006 issue of the journal *Science* concluded that unless humanity makes enormous changes in the way they live and in what they eat, the entire populations of the world's fished species will collapse by about 2048. The study considered evidence from all of the world's 64 large marine ecosystems. They found 91 percent of native species suffered from a 50 percent decrease, and 7 percent were extinct. Continued overfishing as well as coastal land development, habitat destruction, and world pollution are to blame. The study pointed out that nearly 29 percent of species fished have collapsed (defined by being below 10 percent of historic highs). The study authors concluded the fish populations were rapidly decreasing and losing entire functional groups. The study says the oceans will not be able to recover from the decline of so many species. The study authors wrote, "Our analyses suggest that business as usual would foreshadow serious threats to global food security, coastal water quality, and ecosystem stability, affecting current and future generations." Many scientists living in countries throughout the world voiced their opinions in agreement with the study.

We're going to run out of species

"There's no question if we close our eyes and pretend it's all okay, it will continue along the same trajectory. Eventually, we're going to run out of species."

– Dalhousie University marine biologist Boris Worm, leader of the global research team that authored the study, *Impacts of Biodiversity Loss on Ocean Ecosystem Services*; Nov. 2006

Marine biodiversity is weakening

"The loss of marine biodiversity is weakening the ocean ecosystem and its ability to withstand disturbances, to adapt to climate change, and to play its role as a global ecological and climate regulator.

The ocean is home to millions of species. The health of the oceans is strongly dependent upon this marine biodiversity. Life in the ocean is an essential component of climate regulation. Climate change due to human activity has a direct impact on marine species. It alters their abundance, diversity, and distribution. Their feeding, development, and breeding, as well as the relationship between species are affected

Rising temperatures lead to different behavior patterns according to species. Some adapt to temperature changes, while others migrate toward the poles, or to new areas. Other species disappear, as has been observed for certain corals that can rapidly bleach and die when their symbiosis with the unicellular algae, that they shelter and feed on, is interrupted.

The ocean acidification, caused by an increasing absorption of atmospheric carbon dioxide, has a direct impact on the marine organisms with calcareous skeletons or shells: these include phytoplankton, crustaceans, and mollusks.

Extreme climatic events deplete natural environments, for example by erosion and flooding. They disturb marine life in coastal areas, particularly in certain coastal habitats such as mangroves and seagrass beds, which are vital breeding grounds as well as potential CO_2 capture zones."

– Ocean & Climate Platform, ocean-climate.org

The scam of The Marine Stewardship Council

The efforts being made to make seafood eaters more aware of where and how their meal was caught have been proven fraudulent. The Marine Stewardship Council works to certify some types of seafood as sustainable by allowing packaging to feature a blue and white certification label. It's a scam. It's a money-making venture. The MSC also encourages people to avoid eating large fish, and instead to eat low on the food chain by eating oysters, scallops, crabs, and squid, which reproduce more quickly than large fish. However, when it is taken into consideration that the smaller forms of sea life play a part in the health of the seas, nutrient distribution, and help regulate the climate, and have a broader impact out into the land of plants and animals, it should be clear to see why smaller forms of waterlife need to be protected from harm.

Get past the absurdity that certain fish are "safe to eat," or are "sustainably caught."

End the madness of the industrial fishing industry

People do not need to eat seafood. I can understand if they live on an island, or in extreme weather conditions, where other food is not available, or cannot be grown, or wildharvested.

Fish should not be taken from the waters of the world to make pet food, and especially not for food to feed farm animals or farmed fish, which also are so unnecessary for human consumption, and damaging to the climate and environment.

The fish and shelled creatures and coral and marine mammals and sea birds should be left alone and protected from harm.

"Basically, we should stop doing those things that are destructive to the environment, other creatures, and ourselves, and figure out new ways of existing."
– Moby

Please, don't kill or eat sea creatures.

"The fate of animals is of greater importance to me than the fear of appearing ridiculous; it is indissolubly connected with the fate of man."
– Emile Zola

"If you are neutral in situations of injustice, you have chosen the side of the oppressor."
– Desmond Tutu

Livestock by the Billions:
A Global Environmental Disaster

Billions of farmed animals, bred and fed to be dead

"Approximately 70 billion farm animals are reared for food in the world each year.

Approximately two out of every three farm animals in the world are reared on factory farms. In the U.S. about 99 percent of farmed animals are reared on a factory farm."
– AnimalMatters.org

The above figure is only about land animals, including cows, bulls, pigs, goats, chickens, turkeys, ducks, ostriches, and other farmed animals. It doesn't include fish or crustaceans.

Somewhere around two trillion fish are killed every year for food, including to feed farmed fish, and to feed humans and farmed mammals.

"Each day approximately 160 million farmed animals throughout the world are transported to a slaughterhouse."
– Farm Animal Rights Movement

"An average person living in a developed country who is not a vegetarian or vegan will consume approximately 7,000 animals during their lifetime."
– Vegetarian Calculator

The above figure depends on what type of animals people are eating. Fish or mammals. Insects, or shell fish. Birds, amphibians, or reptiles.

Or, dogs. Depending on where you live, people are likely to eat a different variety of animals – or none.

The international meat market, dogs included

In addition to all of the farmed animals being bred, raised for meat, killed, and sold in stores and restaurants, and to industrial, commercial, educational, and confinement facilities – including prisons, medical centers, hotels, schools, cruise ships, country clubs, and the military – and served in cafeterias and buffets of corporations, casinos, and countries outside of the U.S. also raise and kill dogs, snails, frogs, and other animals for human consumption.

"There are more than 17,000 dog-meat farms in South Korea."
– Human Society of the U.S.

Pollution from raising, killing, selling, and eating animals

Who even knows what sort of pollution and environmental damage is done by the animal farming industry in countries outside of the U.S. where other types of animals are raised, fed, killed, processed, and eaten. Humanity does bad enough as is just in the U.S., including by relying on imported meat and dairy that are products of a variety of environmentally ruinous farming practices.

87% of U.S. farmland is used to raise animals for food

"Of all agricultural land in the United States, 87 percent is used to raise animals for food. Twenty thousand pounds of potatoes can be grown on one acre of land, but only 165 pounds of beef can be produced in the same space.

[South and Central American] Rainforests are being destroyed at a rate of 125,000 square miles per year to create space to raise animals for food. Fifty-five square feet of land are consumed for every quarter-pound fast food burger made of rainforest beef."
– PETA, VegNow.Com

Manure

"Nationwide, factory-farmed animals produce 130 times more manure than the human population – the equivalent of five tons of manure for every U.S. citizen.

U.S. factory farms generate more than 350 million tons of manure each year."
– Farm Sanctuary, VegForLife.Org

Ecological damage of feed crop and meat industry

In addition to the incapacitating degenerative diseases, the food poisoning, and the suffering issues of raising, killing, and eating billions of animals throughout the world, there are also the issues of the ecological damage to the planet caused by the massive and multinational livestock industry.

"Cattle and sheep grazing is ecologically destructive and an abomination against our national park system in areas as pristine as the Grand Canyon Park.

Grazing causes rapid depletion of wooded areas by clearing, cultivating, and eroding the soil. Soil losses are as high as 44 tons per acre annually on steep slopes. Woodlands, waterways, and wildlife habitats have been significantly reduced or eradicated entirely due to overgrazing."
– FarmSanctuary.Org

"Most wars are fought over control of natural resources: land, water, oil, and minerals. Yet, animal agriculture is by far the largest user and despoiler of natural resources."
– Citizens for Healthy Options in Children's Education

The livestock and feed crop industries and companies that exist to service them, to ship, and/or package, sell, and cook their products use up more land and resources and create more pollution and environmental damage on the planet than any industry.

Animal farming ravaging biodiversity

At any time in recent years, there are over 34 billion chickens, around 200 million turkeys, over 700 million pigs, over a billion cows, about 500 million other cattle, and hundreds of millions of lambs and other farm animals being raised on millions of acres of land all over the world. This is ravaging the biodiversity of the planet (the full spectrum of living things including plants, fungi, bacteria, other organisms, and insects, birds, mammals, fish, shelled fish, reptiles, and amphibians).

Cow dung pollution altering ecosystems

"Cattle are a chief source of organic pollution; cow dung is poisoning the freshwater lakes, rivers, and streams of the world. Growing herds of cattle are exerting unprecedented pressure on the carrying capacity of natural ecosystems, edging entire species of wildlife to the brink of extinction. Cattle are a growing source of global warming, and their increasing numbers now threaten the very dynamic of the biosphere."

– Jeremy Rifkin, in his book *Beyond Beef: The Rise and Fall of the Cattle Culture*

Benefits of going plant-based

"If everyone in the U.S. ate no meat or cheese just one day a week over one year it would be like not driving 91 billion miles, or taking 76 million cars off the road."

– Environmental Working Group, July 2011

"The less animal-based food you eat, and the more you replace those calories with plant-based food, the better off you are, in terms of your health as well as your contributions to the health of the planet."

– Gidon Eshel, assistant professor of geophysics, University of Chicago, co-author of a study concluding that becoming a vegan does more to reduce greenhouse gases than the type of car you drive. *Earth Interactions* Journal, April 2006; It's better to green your diet than your car, *New Scientist Magazine*, Dec., 2005

The raising of livestock for human consumption and all of the environmentally destructive chemicals used by the animal-farming industry, and all of the support industries and resources needed to support an industry that yearly kills hundreds of millions of cows, over a billion pigs, tens of billions of chickens, over a hundred million turkeys, and millions of other farm animals to sell as meat, causes massive damage to the delicate ecosystems throughout the world.

Livestock in the U.S. produces more than 130 times the amount of bodily waste of the nation's human population.

Meat-rich diets place incredible strain on resources

"With more than nine billion (land) animals raised and slaughtered for human consumption each year in the U.S. alone, modern animal agriculture puts an incredible strain on natural resources such as land, water, and fossil fuels.

More than ever before, people around the world are feeling the effects of the global climate crisis. Many factors play a role in human-induced climate change, including animal agriculture which is responsible for 14.5% of our greenhouse gas emissions worldwide. The United Nations states that animal agriculture 'is exerting mounting pressure on the world's natural resources' and contributes to land degradation, species loss, and water pollution and waste.

Raising billions of animals for slaughter each year also means production of an exorbitant amount of waste. Factory farms in the

United States produce more than 500 million tons of manure annually, according to the Environmental Protection Agency."
– Farm Sanctuary, FarmSanctuary.org

The SAD diet: Standard American Diet

As more people around the world are converting to an American diet style of meat and junk food, expansive stretches of virgin land and land that had been used to grow food for human consumption are being converted to provide space for cattle grazing and for growing feed crops for livestock.

Greenhouse gasses, resources, environmental impacts

"Different meats affect our health and environment differently. Lamb, beef, cheese, and pork generate the most greenhouse gases. They also tend to be higher in fat and have the worst environmental impacts, because producing them uses the most resources – mainly feed, chemical fertilizer, fuel, pesticides, and water. Lamb has the greatest impact. Beef is second. Cheese is third. Beef has more than twice the emissions of pork, nearly four times more than chicken, and more than 13 times as much as vegetable proteins such as beans, lentils, and tofu."
– The Environmental Working Group, July 2011

The tofu myth

Not all vegetarians eat tofu. I don't. That vegetarians eat tofu is something often assumed. When I say I don't eat tofu, I've been asked, "Then, what do you do for protein?" ALL fruits, vegetables, nuts, seeds, sprouts, and seaweeds contain the amino acids our bodies need for creating protein. Tofu isn't needed.

Water use and beef

"For now, let's stay with the original hamburger in its classical form (a meat patty of 100 grams of beef). To produce the 200-gram bun, around 200 liters of water are needed. In Germany, the production of the 100-gram meat patty requires, depending on the type of animal husbandry practiced, between 3500 and 7000 liters of green water (water feeding a crop). Extensive pasturing in South America requires even more water. This example shows more than any other how lifestyle dominates the use of the Earth's green water resources. So, the next time you eat hamburger, imagine the thirty-five virtual bathtubs filled with water you are consuming."
– Wolfram Mauser, *Water Resources: Efficient, Sustainable, and Equitable Use*

Most corn is fed to farmed animals

According to the National Corn Growers Association, in 2011 about 80% of the corn grown in the U.S. was being grown to feed foreign and domestic livestock, including cows, pigs, lambs, bison, chickens, turkeys, and farmed fish.

Exporting feed crops, increasing drought

During the week I'm writing this paragraph in the summer of 2023, there are stories about how much water is being used in Arizona to grow feed crops for dairy cows, veal calves (baby bulls), and other cattle being raised over 8,000 miles away in the Middle East.

A major public employee retirement fund in Arizona invested in a company in New York that bought massive amounts of land in Arizona, which leased the land to grow feed crops for farmed animals in the Middle East.

People in Arizona are on a retirement plan helping to ruin the environment in Arizona. It's a desert region, not natural for growing heavily water-dependent feed crops.

The water used to grow the feed crops has led to a drop in the levels of underground water. This has resulted in people having to spend thousands of dollars to dig deeper wells. People who can't afford it have been increasing reliant on bottled water, and water brought in on trucks to store in cisterns.

Exporting feed crops & dairy equates to exporting water

Feed crops contain water. With Arizona growing feed crops to export to the Middle East, it is also exporting billions of gallons of water contained in those feed crops. This process has worsened drought conditions in the region. It only gets worse, including because more and more land is being used to grow feed crops that are exported.

Arizona is only one of many states where feed crops are grown and then exported to other countries.

California is one of those states, allowing massive quantities of feed crops to be exported, which means billions of gallons of water are being exported from a state regularly experiencing extreme drought. It's also a state where massive amounts of water are used to hydrate cows used to produce milk, cheese, butter, cream, and other dairy products. Large amounts of those products are also exported to other states and countries.

All of this dairy and feed crop production in and export from drought-prone Western states goes on year after year. As the environment suffers.

Nature destruction caused by feed crops, dairy, and meat

On every continent, and on many island nations, the feed crop, dairy, and meat industries cause environmental destruction.

Rainforest destruction by cattle and feed crop industries

"Rainforests cover less than two percent of the Earth's surface, yet they are home to nearly half of our planet's living creatures. Butterflies and birds fill the air, their colors are so intense, no artist could ever match them. Noble jaguars, howler monkeys, vines, fish, gorillas, orchids, lizards, and orangutans flourish there – and nowhere else on Earth.

A four-square mile area of rainforest teems with colorful variety: 750 types of trees, 1,500 different flowers, 125 mammal species, and 400 kinds of birds.

But rainforests are disappearing at the rate of a football field every second. The destruction of the Earth's most ancient complex ecosystem threatens the very survival of the human species.

Over 99 percent of the rainforest species have not yet been studied for possible medical use. A plant that holds the cure for AIDS – or a future epidemic – may be growing somewhere in the rainforest. Or a bulldozer may be crushing the last one right now as you read this.

Rice, potatoes, bananas, chocolate, coffee, oranges, tomatoes, yams and dozens of other food crops originated in the rainforests. The wild strains still found there provide genetic material necessary to keep world agriculture stocks hardy and healthy. Undiscovered rainforest species could provide new sources of food in the future.

Once the rainforest is gone, the vanished species won't return. And once the proud, ancient indigenous cultures are destroyed, the knowledge they possess – knowledge that could benefit all of humanity – will be lost forever."
– The Rainforest Action Network, RAN.Org

Regional extinction of jaguars caused by meat industry

One animal with a range that went from what is currently the Southern United States all through Central America, and down into South America is the jaguar. Jaguars are an apex predator, which means they don't have a natural predator in the wild. But, they have been no match for bullets.

Because of the livestock industry, as of 2023, there is only one known jaguar living in the U.S. and that is in mountains near Tucson, Arizona. Elsewhere, jaguars continue to be killed to protect livestock, and are also in decline because people kill them for their skin that can be sold into the international market, then people can make clothing and furniture upholstered with jaguar skin. It's disgusting.

One region the jaguars have been able to survive, and often without human interference, is the rainforest areas of South America. That has been changing as the rainforests become increasingly fragmented by the lumber industry, the cattle grazing industry, the feed crop industry, coffee plantations, oil palm plantations, and roads.

Rainforests, the lungs of Earth

The Amazon Rainforest is often referred to as the "lungs of the Earth." Of course, all forests and plants contribute to the production of oxygen, especially the plants that grow in the rivers, lakes, and seas. But, perhaps because the Amazon Rainforest is unrivaled in the number of plants and animals it contains, it is often the focus of how humanity is destroying the planet by endangering and eliminating such a wide variety of life forms.

Rainforests being destroyed

Rainforests around the planet are being destroyed by various industries, including the lumber, flower, petroleum, oil palm, coffee, feed crop, pineapple, banana, and beef industries, and the hunting of wild animals, which is sickly called "bushmeat." In many areas, these industries feed off each other. Roads built into the rainforests by the lumber and petroleum industries are then used by hunters and homesteaders. As the awareness grows that there are ways to exploit the wildlife in the forests, people kill the animals to sell as bushmeat, and for their skins, and other body parts. They capture animals to sell into the underground, zoo, circus, and "amusement park" markets. They take rare plants to sell into the floral and houseplant market. And they take whatever they can from the forest to make money.

Not only are native peoples getting in on the game, but also people working for the other industries make money on the side as they learn to sell rainforest life into the underground markets.

Homesteaders move in and destroy more land as they work to clear forest, sometimes only living there long enough to clear forest for cattle grazing or growing feed for livestock, or growing oil palms, coffee, and other crops to be exported to wealthier countries. The wildlife dependent on those forests are killed, or scammer into other areas of the forest. As the feed crop fields and cattle grazing ranches expand, more forestland is cleared, land is degraded, and rivers are polluted, killing fish, amphibians, and other life dependent on the rivers, and larger animals are killed, die off, or relocate. In the whole mess of it all, wildlife diversity is diminished, rare forms of life become extinct, the wildlife food chain is disturbed, and the web of life is altered and degraded.

Bushmeat also destroying rainforests

The growing consumption of bushmeat is one of the most destructive things happening in the rainforests.

Exotic animals formerly hunted by native peoples are being hunted and sold commercially in a growing number of cities and towns in not only the rainforest areas, but more and more in distant cities.

Meat of rainforest mammals, fish, and birds has been found being sold in markets in Paris, Rio, and Asian cities – often as "delicacies."

Mammals, birds, fish, amphibians, and reptiles play important roles in the rainforest. Their normal behaviors pollinate plants, they spread seeds, nutrients, and helpful bacteria and fungi, and their droppings and decomposing bodies fertilize the plants.

As more rainforest animals are killed, the rainforests degrade, nutrient spread decreases, less pollen and fewer seeds are spread, and fewer varieties of plants are common.

Medicines and foods from the rainforests

Many varieties of plants and parts or substances from animals from the Amazon have been used in cosmetics, in pharmaceutical drugs, in other forms of medicine, as unproved cures for various conditions, and as recreational, mind-altering substances.

The rainforest is so diverse in life that many types of plants and smaller animals, and even mushrooms, have not been charted. It is easy to conclude that many unidentified life forms of the rainforests are no longer, as they have been sent into extinction by human behavior – and especially so in the past century.

Because such massive destruction is taking place in the Amazon, it is often said humanity may be permanently losing its access to plant chemicals that could be a cure for some current or future health issue.

Amazon is at tipping point, and possibly has passed it

"Scientists warn that the Amazon is approaching a tipping point beyond which it would begin to transition from a lush tropical forest into a dry, degraded savanna. This point may be reached when 25% of the forest is lost.

In a newly released report, Monitoring of the Andean Amazon Project (MAAP) estimates that 13.2% of the original Amazon forest biome has been lost due to deforestation and other causes.

This equates to more than 85 million hectares (211 million acres), an area about one-tenth the size of the United States or China."
– Liz Kimbrough, *How close is the Amazon tipping point? Forest loss in the east changes the equation*; News.MongoBay.com, Sept. 2022

Millions of acres of the Amazon rainforest have been decimated in the past several decades. In 2010 the estimated loss of Amazon rainforest to clearcutting and burning was 2,500 square miles. The degradation of the rainforest has grandly increased since 2010.

Stupidity of this civilization: addiction to everything

"The biggest stupidity of this civilization, and of all civilizations, was the belief that they will never run out of raw materials – and that these were never in fact 'raw,' but vital, integral pieces of a functioning planet.

Our entire civilization including our cultural norms, way of life, economy, and business are based on strengthening and monetizing addiction to everything.

Unless they stop, all addicts eventually die."
– George Tsakraklides, biologist and author of *In The Grip of Necrocapitalism: The Making and Breaking of A Psychonomy*

The most biodiverse ecosystems on Earth

"Rainforests are the most biodiverse ecosystems of land on Earth, meaning they have more species of animals, plants, fungi, and other life than any other ecosystem. Rainforests are home to millions of species, including 70% of Earth's land animals and plants, and half of Earth's total species, including the ones that live in the ocean. This includes chimpanzees, orangutans, many species of monkeys, rare river dolphins, sloths, gorgeous parrots and toucans, and large and spectacular butterflies.

Our closest relatives, the great apes, live in tropical rainforests and cloud forests. Chimpanzees and orangutans live in rainforests, while gorillas live in cloud forests.

The Bonobo Chimpanzee is the closest relative of the human. Ninety-eight percent of their DNA is identical to ours. Bonobos are a peaceful species of the rainforests that settle disputes without violence. They are threatened with extinction because their rainforest habitat is threatened – as is true of all great apes.

Most of the rainforest species are in the canopy, high up in the tops of the trees. A study of diversity in a rainforest in Peru caused the estimate of the number of species of animals in the world to be revised from 5 to 10 million up to 30 to 60 million.

Costa Rica is smaller than the state of West Virginia, with its widest point being only 185 miles across. Yet has as many bird species as all of the United States and Canada combined. It contains approximately half a million species of animals and plants.

The Amazon River has as many species of fish as the entire Atlantic Ocean.

Rainforests provide a habitat for migratory birds to live during the harsh winters of the temperate and polar regions of the Northern and Southern Hemispheres."

– World Rainforest Fund, WorldRainforest.org

Incinerating the rainforests

"Over 200,000 acres of rainforest are burned every day. That is over 150 acres lost every minute of every day, and 78 million acres are lost every year."

– Leslie Taylor, *Saving the Rainforests*, The Raintree Group

Why are the rainforests being destroyed?

Why are so much of the rainforests being destroyed? Chiefly: It is some branch of the feed crop and livestock industries. Then, palm oil, sun-grown coffee, and other crops, and lumber.

Crops grown on land cleared of rainforest

Rainforest is cleared to supply land to graze cattle and to grow soy, corn, alfalfa, and other crops used to feed the world's billions of farmed animals. (Much of the corn is also used to produce ethanol, even though hemp being grown for cellulosic ethanol would be a better choice – including for the environment.) More and more land is being used to grow oil palm plantations, and coffee plantations.

Second to the livestock industry in the destruction being done to the rainforest is the logging that has been taking place there. Much of the logging being done is illegal, but is so profitable that companies and individuals are willing to take the risks of cutting down rainforest to sell lumber into the international market – and especially into what is shipped to richer countries.

The importance of the natural rainforest systems

"Trees carry out photosynthesis, which produces oxygen and removes carbon dioxide – the main greenhouse gas – from the atmosphere. But when they are cut or burned, trees can no longer perform photosynthesis, so cannot remove carbon dioxide from the air. Instead, burning or dead and decaying trees release the carbon dioxide stored in their trunks into the atmosphere, exacerbating the greenhouse effect.

Rainforests are often destroyed by flooding from huge megadams. In this case, the trees decay under water in the absence of oxygen, which means they release methane instead of carbon dioxide. This is much worse for Earth's climate, because methane is 25 times as powerful a greenhouse gas as carbon dioxide. Thus, destroying rainforests makes human-induced climate much worse, heating the Earth.

Destruction of tropical rainforests accounts for about 17% of global carbon emissions.

The climate crisis cannot be solved without stopping the destruction of rainforests. Without a solution to the crisis of human-induced climate disruption, there will be a tremendous catastrophe greater than humankind has ever experienced, with crop failures and mass starvation, unprecedented shortages of drinkable water, global pandemics, coastlines going under water, record high temperatures, huge droughts and floods, and many other catastrophes.

There is a vicious cycle. Human-induced climate change is rapidly destroying rainforests worldwide. The increased heat from greenhouse gasses people produce cause droughts in rainforests. Hotter air can hold more moisture without dropping it as rain than colder air can. So the warmer air over the forest does not lose its water as rain, but moves away from the equator, and drops rain in temperate regions, where it is too cold for the air to hold its moisture. Thus, rainforests are experiencing droughts all over the Earth.

Rainforests normally have large amounts of rain. This keeps fires in the rainforests small and limited in size. But when rainforests are dry from droughts, fires are more easily started and can burn much larger areas.

Since the droughts in rainforests from global warming have appeared, there have been massive fires in rainforests in Mexico, Brazil, Africa, Indonesia, and other rainforests, destroying tremendous areas of these forests.

The Indonesian fires have causes massive air pollution in Southeast Asia. The fires are often started by land developers, ranchers, and corporate farmers. The droughts also cause the edges of rainforests to die. After the trees on the edges die and fall over, there is a new edge whose trees then dry out and die, in an endless cycle that can continue until the entire rainforest is gone.

Rainforests are important in the regulation of the global water cycle. Thus, the destruction of rainforests affects Earth's water cycle and causes floods and droughts far from rainforests.

Sao Paulo, Brazil's largest city, is suffering from serious droughts due to the destruction of the Amazon rainforest. This megacity of over 12 million people is literally running out of water.

Deforestation in the Amazon and Central America severely reduces rainfall in the lower U.S. Midwest during the spring and summer seasons, and in the upper U.S. Midwest during the spring, and respectively deforestation of Southeast Asia affects rainfall in China and the Balkan Peninsula, significantly.

High-resolution simulation of destruction of the Amazon rainforests showed 10-20% less rainfall for the coastal northwest US and the Sierra Nevada, and declines of snowpack in the Sierra Nevada of up to 50%.

Rainforests also play an important role in the local water cycle. Rainforest trees add water vapor to the atmosphere by transporting water in the ground through their roots, up their trunks, and through their leaves into the atmosphere in a process called evapotranspiration. This causes rain.

Half of the rain in the Amazon rainforests is caused by the trees there, while the other half comes from water evaporated by the sun in the Atlantic Ocean.

The rain that rainforests produce is needed by local farmers to grow food, people and animals for drinking, trees and plants to grow, and the rivers and lakes of the forest. These rivers and lakes support fish, freshwater dolphins, turtles, and invertebrates, such as insect larvae and shrimp.

Rainforest water is purified by tree roots and fungi underground, where toxins are removed. Roots of rainforests trees store large amounts of water. Without the rainforests to soak up rain and release it slowly, floods and droughts become more common. In the Amazon rainforest, more than half the water in the ecosystem is stored within the plants.

Trees shade rivers and keep them cool enough for fish and other animals and plants in them to thrive. They stabilize river banks, and prevent their erosion. Without this, sediment from soil would pollute rivers and lakes in the forest, and many fish, river dolphins, and aquatic invertebrates would die.

Rainforest trees shade and protect the soil, blocking the hot, intense tropical sun from hitting it during the day. They hold heat in at night. Destroying the forests leads to more extreme temperature swings that are harmful to plants and animals. The trees also protect the soil from intense tropical rains. When the trees are removed, the rain washes away the soil, and the sun bakes the soil into a hard, brick-like state. In a short time, this can result in a desert on which no trees or food crops can grow.

Rainforests affect the reflectivity of sunlight by the Earth. This is called the Earth's albedo. When they are cut, the albedo changes, and sunlight is reflected differently. This changes wind patterns and therefore rain patterns of the planet. This causes floods and droughts, reducing agriculture yields, and the ability to obtain drinking water. The floods cause soil erosion, making it harder for wild plants and trees to grow, and for humans to grow food."
– World Rainforest Fund

Chico Mendes and protecting the Brazilian rainforest

Look into the tragic story of a Brazilian rubber tapper and union leader Chico Mendes who was murdered by cattle ranchers. It adds another layer to how much damage the corrupt cattle and feed crop industries have caused to the life-sustaining rainforests.

Besides for legal and illegal lumber operations that decimate forests, the main reason rainforests in South and Central America have been destroyed by cutting and burning is largely to clear new land for cattle grazing and to plant feed crops for farmed animals.

When rainforest is cut down by the lumber industry, much of it ends up as land to graze cattle, and to grow feed crops – and also to plant corn to make ethanol, plant sun-grown coffee, and plant more and more oil palms to produce palm oil – which has its own set of environmental damage and health issues.

Corporate greed fuels environmental destruction

"Large corporations make a great deal of money by cutting and destroying very large areas of rainforest to plant cash crops to export to countries such as the U.S., and those in Europe. The forests are cut down to create land for vast plantations, where products such as bananas, palm oil, pineapple, sugar cane, tea, and coffee are grown. Just as with local farming, the nutrients are depleted within a few years, and the company moves on to cut another large area of rainforest, destroying it forever. But in this case, the area of rainforest destroyed per crop planted is much larger than when small farmers burn the forest to plant crops.

In large corporate farms, herbicides and pesticides are used in great amounts, because tropical rainforest has so many species of plants that could compete with the cash crops and insects that could eat them. These toxins kill rainforest animals on land, in rivers, and in lakes, and cause cancer and other diseases in the local people. The same principal applies as that for slash and burn agriculture. The nutrients in the soil to grow the cash crops are depleted in a few years, and the agricultural corporations must cut more rainforest to start raising their crops in a new area (to maintain the same output, maintain company profits, and satisfy stockholders).

Many rainforests in Central and South America have been burnt down to grow grass for cattle to graze on. Cattle ranching in the rainforests supplies cheap beef (primarily for export to richer countries). It is often used to make hamburgers for fast food restaurant chains.

Cattle ranching in the rainforest is spectacularly inefficient. It is estimated that for each pound of beef produced, 200 square feet of rainforest is destroyed. Incredibly, another estimate is that it takes 5 acres of grassland from cleared rainforest in the tropics to support one cow.

Large amounts of land and water are used by the beef and feed crop industries. The rainforest land ends up as ruined. The methods to raise the cattle are often inhumane. In addition, eating beef causes heart disease, clogged arteries, some cancers, autoimmune disorders, and other chronic and degenerative diseases in humans. As in agriculture, the nutrients in the soil to grow the grass get depleted in a few years, and the ranchers cut more rainforest to raise cattle in a new area.

When certain animals are killed off, many other species of animals and plants can die out. For example, if the jaguar is sent extinct locally in a rainforest, its prey will have population explosions. They will then eat so much of the animals and plants that they normally eat that these will go locally extinct, or greatly decrease in number. Then animals and plants dependent on them will go locally extinct in a catastrophic chain of die offs throughout the food webs of the rainforest.

Approximately 65-70% of deforestation in the Amazon occurs because rainforest is cut to make room for cattle ranching and feed crops, not cutting trees for wood. Worldwide, demand for beef is increasing, because so many people eat so much of it. Large areas of the rainforest are being destroyed so that ranchers can meet this demand. Even if you eat only beef grown outside of rainforest areas, you are still contributing to the problem by raising prices and demand, encouraging more Brazilian farmers to raise cattle."
    – World Rainforest Fund

Clearcutting kills

Clearcutting kills off many kinds of plants, insects, mammals, amphibians, reptiles, birds, and beneficial soil organisms, and also destroys natural water filtration systems within the rainforests.

It's to satisfy the world market of greed

The cattle raised on the defiled rainforest land are used to supply beef to the Orient, to the Middle East, to North America, and to Europe – where millions of acres of land are already ruined to raise, feed, and grow feed for hundreds of millions of cattle – and tremendous numbers of other farmed animals. The destruction is deliberate. This is ecocide.

Farmed animals eat more food than humans

"The livestock population of the United States today consumes enough grain and soybeans to feed over five times the entire human population of the country. We feed these animals over 80 percent of the corn we grow, and over 95 percent of the oats.

Less than half the harvested agricultural acreage in the United States is used to grow food for people. Most of it is used to grow livestock feed. This is a drastically inefficient use of our acreage.

For every sixteen pounds of grain and soybeans fed to beef cattle, we get back only one pound as meat on our plates. The other fifteen are inaccessible to us. Most of it is turned into manure."
– John Robbins, in his book *Diet for a New America*

Trout population decline caused by animal farming

In 1994 the U.S. Fish and Wildlife Service announced California's state fish, the golden trout found in the waters in the High Sierras of California, was a candidate for the endangered species list. The declining population of the fish was blamed on cattle grazing in the U.S. government-owned Golden Trout Wilderness land adjacent to the streams where the fish spawn. The cattle the government allows ranchers to graze there have eroded the land and widened the riverbanks. As the cattle trample the soil it becomes compacted, making it unsuitable for new plant growth and preventing rainwater from being absorbed. Instead, the rainwater runoff carries soil into rivers and streams. The plants that do take up root are often invasive weeds. The vegetation degradation, soil erosion, and the manure from the cattle has affected the water temperature and clarity, smothered fish eggs, and has had a negative impact on the population of insects, which are eaten by fish, birds, bats, some reptiles, and amphibians.

175

Cows need more water than crops grown for humans

"In California, the number of gallons of water needed to produce one edible pound of: wheat: 25 gallons; beef: 5,214 gallons.
Energy expended to produce one pound of grain-fed beef: equivalent to one gallon of gasoline."
– EarthSave.Org

"McDonald's equals slavery and starvation."
– Graffiti on a wall in Quito, Ecuador

Meat industry: half of water & 33 percent of raw materials

"More than half of all the water and 33 percent of all the raw materials used for all purposes in the United States are used in meat production... the average chicken-processing plant may use 100 million gallons of water daily. More than 260 million acres of U.S. forest have been cleared to grow crops to feed to cattle. Cattle-grazing in the western United States has led to soil erosion and desertification."
– *Take a Step Toward Compassionate Living*, by the People for the Ethical Treatment of Animals, PETA-Online.Org; 2004

Massive quantities of chemicals used by meat industry

"Most U.S. livestock are fattened on soybean meal, corn, and other feed crops that require large amounts of fertilizer, fuel, pesticides, and water. It takes 149 million acres of cropland, 167 million pounds of pesticides, and 17 billion pounds of nitrogen fertilizer to grow this much feed every year. When fertilizer is spread on soil, it generates nitrous oxide (N_2O), which has 300 times the warming effect of CO_2. Pesticides and fertilizers often end up in field runoff, polluting rivers, ground water, and ultimately, the ocean (like the Gulf of Mexico's famous 'dead zone'). Feed crops are heavily subsidized by taxpayers in the federal farm bill – to the tune of $45 billion over the past 10 years."
– Environmental Working Group, July 2011

Meat industry is not energy efficient: Massively wasteful

Raising livestock and growing crops to feed livestock is not energy efficient. It takes significantly larger amounts of land, water, and other resources to produce cattle, poultry, and hog meat than it does to produce fruits, vegetables, herbs, legumes, and grains for human consumption. More than half of the water used in the U.S. goes to grow feed crops and to provide water for livestock. Those uses of water have been partially blamed for the droughts in California where water consumption by the livestock industry exceeds that of the state's human population.

What is needed for a greener society

"Although eating less meat is important for improving our personal health and living more gently on Earth, that alone won't stop climate change or environmental damage. Even if everyone in the U.S. chose a vegetarian diet – the equivalent of taking 46 million cars off the road – it would make only a small dent in overall emissions. To bring them down significantly, we need to convince our elected officials to enact a comprehensive energy and climate policy that puts the nation on a path to green energy. We also need better policies and stronger regulatory enforcement to reduce meat production's negative impact on soil, air, water, and animal welfare."
– The Environmental Working Group, July 2011

Droughts in other regions of the planet, such as Australia, Africa, and Europe have also been directly related to animal farming and growing massive quantities of food for farmed animals.

All of this is mostly for raising cattle, a species native to Turkey and Pakistan. Because of humans, they are an invasive ecocide species.

47% of California's usable water goes to animal ag

"In his piece on the need to fallow or re-purpose San Joaquin Valley farmland, columnist George Skelton ignores the fact an estimated 47% of all usable water in California goes toward animal agriculture.

Skelton says we need to 'plant fewer thirsty crops, such as almonds.' He ignores the egregious use of lower Colorado River water for alfalfa, which is grown in the California desert, and much of it sent overseas to feed livestock.

In Southern California, residents are asked to limit water use to 500-600 gallons per week. One decent-sized hamburger requires more than 600 gallons to produce.

A 10% reduction in animal food production would achieve a 4.7% reduction in total water use. It would also substantially increase water available to California households, and result in significant greenhouse gas reduction."
– Booker Pearson, Letters to the Editor: Using Animals for Food is Draining California's Water Supply; *Los Angeles Times*; Feb. 2023

Alfalfa grown in the US drought feeds Arabian cattle

"Four hours east of Los Angeles, in a drought-stricken area of a drought-afflicted state, is a small town called Blythe where alfalfa is king. More than half of the town's 94,000 acres are busy blue-green fields growing the crop.

Massive industrial storehouses line the southern end of town, packed with thousands upon thousands of stacks of alfalfa bales ready to be fed to dairy cows – but not cows in California's Central Valley, or Montana's rangelands.

Instead, the alfalfa will be fed to cows in Saudi Arabia.

The storehouses belong to Fondomonte Farms, a subsidiary of the Saudi Arabia-based company Almarai – one of the largest food production companies in the world. The company sells milk, powdered milk, and packaged items such as croissants, strudels, and cupcakes in supermarkets and corner stores throughout the Middle East and North Africa, and in specialty grocers throughout the U.S.

Each month, Fondomonte Farms loads the alfalfa on to hulking metal shipping containers destined to arrive 24 days later at a massive port stationed on the Red Sea, just outside King Abdullah City in Saudi Arabia."

– Lauren Markham, *Who Keeps Buying California's Scarce Water? Saudi Arabia*; TheGuardian.com, March 2019

Saudi Arabia outlawed growing alfalfa because of drought

As the article in the guardian points out, not only is Saudi Arabia largely desert land with limited water, and not a place for feed crops like alfalfa to grow. To save water, the country outlawed alfalfa farms in 2016.

The export of feed crops to places as far away as Saudi Arabia has been going on for decades. It is so common that feed crops are of the main U.S. exports. This is a key part of the global ecocide.

When Saudi Arabia outlawed growing alfalfa on its lands, it spurred a great increase in the import of feed crops from as far away as the U.S. west, and Argentina to feed Saudi Arabia's nearly 100,000 cows.

China's increase in meat consumption

"Despite importing almost no alfalfa until the mid-2000s, China has quickly emerged as the world's largest importer of alfalfa, as well as the largest market for U.S. alfalfa, in recent years. Indeed, China's alfalfa imports increased steadily from 19,601 metric tons in 2008 to 1.38 million metric tons in 2018, and the United States accounted for an average 97.01% of China's alfalfa imports from 2007 to 2017.

The significant increase in China's imports of alfalfa and many other agricultural products like soybeans, powdered milk, and pork has been very helpful for U.S. agriculture, as many U.S. farm sectors are highly dependent on the export markets. Increased agricultural exports to China have not only increased revenue for many U.S. farmers and agribusinesses, but have also helped the United States reduce its large trade deficit with China."

– Qingbin Wang, Zou Yang, *China's alfalfa market and imports: Development, trends, and potential impact of the U.S.-China trade dispute and retaliations*; Elsevier: *Journal of Integrative Agriculture*; Science Direct, ScienceDirect.com; 2020

Marketing U.S. dairy and meat in China = greed

To help increase the meat and dairy consumption in China and other countries, in ways benefitting the U.S. meat, dairy, and feed crop industries, the U.S. helps pay for the meat and dairy advertising and marketing in other countries. This continues at a time when it is scientifically established that eating meat and dairy increases a variety of chronic and degenerative diseases, and the meat, dairy, and feed crop industries contribute greatly to climate change, global warming, soil degradation, water pollution, water dead zones, and ocean acidification. What a strange situation it is when countries say they are working to reverse climate change and air and water pollution, and help protect wildlife, while they are also helping to finance the most environmentally destructive practices that are paving the way for extinction of animals, including the human animal. Because of unwise food choices.

"About 1,000,000 acres of alfalfa are irrigated in California. This large acreage coupled with a long growing season make alfalfa the largest agricultural user of water, with annual water applications of 4,000,000 to 5,500,000 acre-feet."

– Irrigating Alfalfa in California with Limited Water Supplies; UCManageDrought.UCDavis.edu

Growing and exporting crops: pollution-intensive

Growing and exporting feed crops is an energy-intensive practice involving massive amounts of water, fuel, equipment, trucks, trains, and highly polluting ships. Massive amounts of chemicals are used on feed crop fields, including corn, soy, alfalfa, oats, and other crops.

Western U.S. drought and feed crops

I don't know what the process does to the land in Argentina, but it can't be good. In the U.S. growing feed crops and hydrating farmed animals has helped cause a plunge in ground water levels, and in lakes and rivers already suffering from climate change. It is water people of all income levels depend on for their hydration, and to grow other crops – and, unwisely, to water lawns and golf courses in dry regions.

To grow the farmed animal feed crops in California, Arizona, Nevada, and Utah, much of the water is coming from the same water

179

sources used by the populations of Southern California. It is water from the Colorado river basin, which stretches from Wyoming down through Utah and Colorado, Arizona, New Mexico, and Nevada, to California and to the edge of the Sea of Cortez in Mexico. The feed crop and dairy industry – and keeping all the lawns and golf courses green in dry lands – is all vastly unsustainable, and environmentally ruinous.

Even the dry state of Nevada lists its leading crop as alfalfa, with the second being hay – to feed mass-bred, thirsty cattle all so people can eat heart-clogging, cancer-inducing, inflammatory beef and dairy.

U.S. citizens reliant on foreign companies growing feed crops

Foreign companies growing feed crops in the U.S. west employ Americans to work the fields, manage the warehouses, and the shipping. The companies also buy feed crops from other farms. The farmers and the workers are dependent on the pay in a completely unsustainable farming operation – to feed cows halfway around the planet.

Some of the employees are drone operators who can monitor the fields, and soil analysts, and those who know the technology of remote watering systems controlled by computer. Those employees are also a branch of modern-day beef and dairy industry.

Water being exported through feed crops, dairy, and meat

It's not bad enough that so much water is being removed from California, Arizona, and other states so other countries can have beef and dairy. The unsustainable practices have been going on for decades, and are increasing drought, environmental degradation, human disease, pollution, climate change, ocean acidification, and extinction.

Massive water used in California for meat industry

The beef industry in California is the state's third-largest agricultural sector. It couldn't exist without massive amounts of water to hydrate the animals, and to grow feed crops.

The number one reason California has had droughts where people were told not to use water and some people were fined for using more of their share of water is because so much water is going to grow feed and hydrate cattle and hogs. The livestock herds in California deplete the water supplies even more during the times where there is no rain and the temperatures are high. The heat increases the amount of water needed to grow the crops, and the water needed to keep the animals hydrated.

Water for farm animals, and to grow their food, is pumped out of the ground, draining aquifers; or is taken from streams, lakes, and rivers, which is ruinous for wildlife populations dependent on the water.

Public lands used to graze cattle, damaging land, water, and wildlife

"Livestock grazing on public lands accounts for less than one-tenth of one percent of employment in the eleven western states, including in Colorado (according to a study by Thomas Powers, chairman of the University of Montana's Department of Economics). However, this activity costs taxpayers anywhere from three to five hundred million dollars per year (according to the Cato Institute). More significantly, cows and sheep on public lands pollute streams and rivers, and jeopardize the continued survival of many rare wildlife species (according to the Congressional General Accounting Office)."
– *Colorado Wolf Tracks*, the newsletter of Sinapu.Org; 1996

Government subsidies support livestock and feed grain industries

The U.S. and other governments spend tremendously to build and manage water systems for irrigating massive fields of feed crops, and for hydrating cattle, hogs, and other farmed animals. The water is polluted by the chemicals used to grow the feed, and by the farm animal urine and feces, and the pharmaceutical and other chemicals used on the farms.

The pollution caused by breeding, feeding, killing, packaging, and cooking billions of animals poisons the land, air, aquifers, streams, rivers, ponds, lakes, marshes, oceans, wildlife, and you.

"The U.S. government spent $22 billion to build 133 water projects in the West. Every year the government spends $7 billion on these water projects. Farmers pay less than $1 billion per year to the government for water."
– NPR, 1997

Tremendous waste of resources to cause this death and destruction

"One pound of beef takes 2,500 gallons of water, eggs 477 gallons, and cheese nearly 900 gallons.

82% of the world's starving children live in countries where food is fed to animals in livestock, and then sold to wealthier countries.

Animal agriculture produces 65% of the world's nitrous oxide emissions, which has a global warming impact 296 times greater than carbon dioxide.

Raising livestock generates nearly 15% of total global greenhouse gas emissions, which is greater than all the transportation emissions combined. It uses nearly 70% of agricultural land, and is the major contributor to deforestation, biodiversity loss, and water pollution."
– *Animal Agriculture Impact*, University of Colorado; March 2023

Use your food choices

"You can use your food choices, your daily habit of eating, to say yes to life in a very profound way, and to say no to the corporate culture destroying the planet, and our communities. You can use your food choices, every meal, every bite, as an opportunity to take a stand for life, to take a stand for compassion."
– John Robbins, *The Food Revolution*

As I state earlier, the human dietary need for animal protein (including from meat, milk, and eggs) is absolute zero. Human health flourishes on a plant-based diet consisting largely of fruits, vegetables, sprouts, nuts, seeds, and seaweeds. Preferably organically-grown, unprocessed, and also much or most of it raw.

Dairy industry: manure, ammonia, methane, and wildland damage

The California dairy industry is a billion-dollar industry that relies on the milk from of a population of about 1.7 million dairy cows. These cattle create more than 65 billion pounds of manure each year. This results in ammonia emissions adding significant amounts of pollution to the environment. Much of the manure is spread on farmland in the surrounding regions. Many dairies produce so much manure it can't be used and it ends up poisoning the land, rivers, lakes, and ocean.

Methane digesters to generate electricity

A small number of California's 1,100 dairies have their manure hauled away by companies that bring it to "methane digester" plants that heat the slurry so bacteria breaks it down to release methane gas to generate electricity. (Search: Methane and global warming.)

Livestock: methane, nitrous dioxide, ammonia, and hydrogen sulfide

"Through their unique digestive process (enteric fermentation), cows, sheep, and other ruminant livestock release substantial amounts of methane – a greenhouse gas 25 times more potent than carbon dioxide (CO_2).

Animal waste causes substantial water and air pollution and emits nitrous dioxide and methane. U.S. livestock in confined feedlots generate about 500 million tons of manure a year, three times the waste produced by the entire human population. Manure is a valuable nutrient for plants, but it can leach pollutants – including antibiotics, metals, nitrogen, and phosphorus – into groundwater when storage facilities leak – or too much is spread on farm fields. Decomposing

waste also releases polluting dust, smog, and toxic gases (ammonia, hydrogen sulfide) that can cause itching and dizziness in people."
– Environmental Working Group, July 2011

"The beef cattle industry generates approximately 24.4 million metric tons of manure per year, dairy farming 19 million tons, poultry 12.7 million tons of litter and manure, and swine production generates manure equivalent of 14.5 million tons."
– Range Cattle Research & Education Center; Feb. 2022

Cows contribute to lung-damaging, particulate-laden smog

Scientists at UC Davis found that cows emit a significant amount of gasses from their chewing and regurgitating. Combined with nitrogen oxide emissions from vehicles and factories, the ammonia and methane from millions of cattle creates particulate-laden smog. This is damaging to lung tissue, and is especially a risk to babies, children, pregnant women, the elderly, people with weakened immune systems, and, in reality, everyone who breathes. Smog also affects plant life as it reduces growth and photosynthesis, and speeds aging within leaves. This has an impact on crop farming as well as on forests, wildlands, and on wildlife of all varieties. The smog is the main reason dairy farming regions of California often have some of the worst air quality of the continent.

"When emissions from land use and land use change are included, the livestock sector accounts for nine percent of CO_2 deriving from human-related activities, but produces a much larger share of even more harmful greenhouse gases."
– Food and Agriculture Organization, Rome, Italy; in a report on worldwide pollution caused by the cattle industry; 2006

According to a 1995 estimate by the San Joaquin Valley Air Pollution Control District, the average dairy cow releases 19.3 pounds of volatile organic compounds every year. California's San Joaquin Valley contains the most dairy cows of any region of the continent, and it also averages out to be the smoggiest air in North America. If you buy milk, cheese, and other dairy products, you support the industry, and play a part in this pollution and related environmental damage.

Animal farming causes more water pollution than any industry

"Raising animals for food causes more water pollution than any other industry in the U.S. because animals raised for food produce one hundred thirty times the excrement of the entire human population. It means 87,000 pounds per second. Much of the waste from factory farms and slaughterhouses flows into streams and rivers, contaminating water sources.

Each vegetarian can save one acre of trees per year. More than 260 million acres of U.S. forests have been cleared to grow crops to feed animals raised for meat. And another acre of trees disappears every eight seconds. The tropical rainforests are also being destroyed to create grazing land for cattle."
– Thich Naht Hanh, *Eating for Peace*, on mindful consumption.

"Raising animals for food consumes more than half of all the water used in the United States. It takes 2,500 gallons of water to produce a pound of meat, but only 25 gallons to produce a pound of wheat. The amount of water used in the production of the meat from an average steer could float a destroyer."
– VegNow.Com

North Carolina's animal farming water pollution problem

North Carolina is one state with massive water pollution issues caused by animal farming. There are millions of chickens, turkeys, and pigs in the state, nearly all kept in metal warehouse "factory farms." These create an enormous amount of urine and feces. There have been numerous problems with the animal waste contaminating creeks, rivers, ponds, lakes, and killing millions of fish and other water life. As mentioned elsewhere in the book, there have been situations of flooding carrying drowned and bloated farmed animals all the way out to sea.

When the storms wash away the animals, they also wash tremendous amounts of animal waste into the water.

Factory farms are often located on cheaper land, especially places where houses and other structures shouldn't be built: floodplains.

Hundreds of millions of farmed birds, millions of pigs = pollution

"North Carolina is overrun with 7,352 swine and poultry factory farms that produce vast quantities of manure.

Of those operations, 156 are in or just outside of floodplains – but even more facilities are probably at a higher risk of flooding.

As the climate emergency intensifies, flooding from severe storms is more likely to flood factory farms, threatening human health and the environment.

In total, we've identified 4,863 poultry facilities in the state that load about 544 million chickens and turkeys into huge, often windowless barns, and 2,489 swine operations cramming almost 8.8 million hogs into tight quarters.

These concentrated animal feeding operations, or CAFOs, generate extraordinary amounts of excrement and urine, which fouls waterways, sullies the air, and sickens people – all with little oversight or operator accountability.

In North Carolina, floodplains are designated by the state's Department of Public Safety working together with the Federal Emergency Management Agency. For most areas, there are 100-year floodplains, and 500-year floodplains: the former supposedly have a 1 percent chance of flooding in a given year, versus a 0.2 percent chance of flooding each year in the latter. But many places across the U.S. in the 100 year floodplain and even the 500-year floodplain are now seeing flooding regularly.

We found CAFOs whose properties were at least partially in a floodplain, or were entirely surrounded by a floodplain, but not technically considered part of it because they were built on slightly elevated ground. We also found swine CAFOs whose enormous cesspools or barns were built immediately adjacent to the 100-year floodplains.

It's important to note that even CAFOs not located in floodplains can flood from heavy rain events.

No one really knows the extent to which CAFO animal waste already contaminates drinking water, especially in parts of the state with a shallow water table. But we do know that the risk multiplies with heavy rains and flooding.

When CAFOs flood, the enormous quantities of manure they produce and store on-site can wash into nearby creeks, and rivers, carrying with it antibiotic-resistant bacteria and pathogens like E. coli and salmonella."

– Sarah Graddy and Al Rabine, *EWG analysis: North Carolina's factory farms are everywhere – even floodplains and other flood-prone areas*; Environmental Working Group; Dec. 2022

Overflowing cesspools

Often when a cesspool overflows on a factory farm, it isn't reported right away. Sometimes they never are. With factory farming becoming more common in other countries, it is a wonder how much damage they are doing to the environment, and how many wild animals have been killed from the pollution – especially fish, salamanders, frogs, and other water-based animals. It also impacts human health in a variety of ways, including increased rates of lung disease, cancers, and other maladies among those who live near or work in or near factory farms.

Slaughterhouses cause massive pollution

"More than 9 billion animals are killed in slaughterhouses across the United States each year – that is, on average, over 17,000 animals per minute. Slaughterhouse byproducts such as fat, bone, blood, and feathers often are sent to animal rendering facilities for conversion into tallow, lard, animal meal, and other products. Both slaughterhouses and rendering facilities require a near-consistent flow of water, and they discharge staggering quantities of dangerous and damaging water pollution.

According to the Environmental Protection Agency, slaughterhouses and rendering facilities are the country's largest industrial source of phosphorus pollution, and second largest industrial source of nitrogen pollution. Nitrogen and phosphorus are naturally occurring nutrients necessary for plant growth. But excessive quantities of nitrogen and phosphorus pollution can accumulate in waterways, rendering water unsafe for drinking and unfit for outdoor recreation. Nitrogen and phosphorus pollution also promote harmful algal blooms, which in turn can cause reductions in oxygen levels as they decompose, resulting in fish kills and other harm to aquatic life.

Not only does water pollution from slaughterhouses and rendering facilities threaten human health and the environment, but it also exacerbates injustice.

We analyzed 184 slaughterhouses and rendering facilities identified by the EPA, all of which discharge pollution directly to rivers and streams, and found that these facilities are disproportionately likely to be located within one mile of communities that the EPA classifies as 'low-income' or 'linguistically isolated.' Similarly, our analysis of 308 slaughterhouses and rendering facilities that discharge pollution indirectly through sewage treatment plants shows these facilities tend to be located within one mile of communities classified as 'low-income' or 'minority.'

To make matters worse, slaughterhouses and rendering facilities are often located near additional slaughterhouses, rendering facilities, and concentrated animal feeding operations, commonly known as CAFOs, compounding the risks they pose. For example, according to the EPA, there are at least 27 indirect-discharging slaughterhouses and rendering facilities located in Vernon, California, where more than 90 percent of the population are people of color."

– Alexic Andiman, Manny Rutinal, Mustafa Saifuddin, *We Sued the EPA to Restrict Water Pollution from Slaughterhouses. And we Won*; EarthJustice.org; May 2023

Slaughterhouses dump millions of pounds of toxins into waterways

"Although slaughterhouses account for a small portion of most meats' overall carbon footprint, they dump millions of pounds of toxic pollutants into America's waterways (nitrogen, phosphorus, ammonia, etc.). Eight slaughterhouses consistently rank among the nation's top 20 industrial polluters, responsible for discharging 30 million pounds of contamination in 2009."

– Environmental Working Group, July 2011

Meat production increasing, failing the environment

According to the United Nations, global meat production in 2001 was 229 million tons. By 2022, global meat production was 340 million tons. It has been expected to be about 400 to 450 tons by the year 2050. However, with the current rate of global environmental degradation, species reductions, political instability, and climate breakdown as I'm writing this in 2023, it appears likely that the global food, energy, transportation, and medical industries, and monetary systems may simply collapse this century – massively reducing human population to a small fraction of what it currently is.

America is world leader in cow milk production

The milk industry produced an estimated 544 million tons of milk in 2022. The U.S. is the world's second largest producer of milk, with about 102 million tons produced in 2022. India is the world leader in milk production, with about 24% of world production – but a significant amount of that milk comes from buffalo. America is the world leader in cow milk production.

Other top ten milk-producing countries are: China (which is heavily reliant on U.S. feed crop imports), Brazil, Germany, Russia, France, New Zealand, Turkey, and the combination of England, Northern Ireland, Wales, and Scotland.

China is increasingly exporting more cow milk to Russia, as Russia imports less from other countries. The U.S. subsidizes the feed crop industry, which exports feed crops to China to feed their dairy cows, and has been increasing its dairy herds as it supplies more and dairy to Russia, which has invaded Ukraine, which is being assisted in their war against Russia by the U.S. Governments function in absurd ways.

Animal protein and human disease

"Adult men, on average, get twice as much protein as they need. In 2009, the U.S. produced 208 pounds of meat per person for domestic consumption, nearly 60 percent more than Europe. The evidence is clear that eating too much meat – particularly red and processed meats – is associated with serious health problems. We can improve our health, shrink our waistlines, and even extend our lives by eating more vegetables, and less meat.

Eating a lot of red and processed meat is linked with increased risk of heart disease, certain cancers, obesity, and, in some studies, diabetes. Two examples:

• A 2009 National Cancer Institute study of 500,000 Americans found that people who ate the most red meat were 20 percent more likely to die of cancer and at least 27 percent more likely to die of heart disease than those who ate the least.

• A seven-year study of almost 200,000 people found that those who ate the most processed meats had a 67 percent higher risk of pancreatic cancer than those who ate little or none."

– Environmental Working Group, July 2011

"When we kill the animals to eat them, they end up killing us because their flesh, which contains cholesterol and saturated fat, was never intended for human beings."

– William C. Roberts, M.D., editor of *The American Journal of Cardiology*

"The awful wrongs and sufferings forced upon the innocent, helpless, faithful animal race form the blackest chapter in the whole world's history."

– Edward Freeman

"Love the world with an all-embracing love. Love the animals; God has given them the rudiments of thought and joy untroubled. Do not trouble them, do not harass them, do not deprive them of their happiness, do not work against God's intent. Man, do not pride yourself on your superiority to them, for they are without sin, and you with your greatness defile the Earth."

– Fyodor Dostoyevsky

Pollution Caused by Farming Animals

Most pollution, and environmental damage linked to diet

Most pollution on this planet is caused by the meat, dairy, and egg industries – including the meat, dairy, and egg packaging, distribution, marketing, cooking, and consumption processes. This is because of the misuse of resources such as fuel, water, land, metal, plastics, paper, electricity, and other materials and resources used to get the tissues of mammals, birds, and water creatures, and also dairy, and eggs onto dining tables.

Consider the following:

Resources used to keep the meat and dairy diet going

1. The materials and fuel used to manufacture the equipment used to farm the food for billions of farm animals.

2. The supplies and fuel used to run and maintain the equipment used to grow the food for billions of farm animals.

A tremendous amount of sea life is being used in farmed animal feed, and in fertilizers used to grow enormous amounts of food for billions of farmed animals.

The fishing industry, and all of the fuel, ships, and equipment used for capturing and/or farming sea creatures used in the form of feed and fertilizer can be included in this.

3. The supplies and fuel used to transfer the food to feed billions of farm animals.

4. The supplies, fuel, and water used to manufacture the equipment used to raise billions of farm animals.

5. The supplies, fuel, and water used to run and maintain the equipment used to raise billions of farm animals.

6. The supplies, fuel, and water used to transfer billions of farm animals to the slaughterhouses.

7. The supplies, fuel, and water used to create and maintain the trucks and other equipment used to transport billions of farm animals to slaughterhouses and processing and packaging facilities.

8. The supplies, fuel, and water used to manufacture the equipment that runs the slaughterhouses where billions of farm animals are killed.

9. The supplies, fuel, and water used to run and maintain the slaughterhouses and processing and packaging facilities.

10. The supplies, fuel, and water used to manufacture the equipment that transfers meat, dairy, and egg products to storage facilities.

11. The supplies, fuel, and water used to run and maintain those processing and storage facilities.

12. The supplies, fuel, and water used to manufacture the equipment used to transfer the meat, dairy, and egg products to the markets.

13. The supplies and fuel used to run and maintain the equipment used to transfer trillions of pounds of refrigerated meat, dairy, and egg products to the markets.

14. The supplies, fuel, and water used to manufacture the equipment used in the markets – from the buildings to the shelving and sanitary maintenance, to the heating, air conditioning, refrigeration units.

15. The supplies, fuel, and water used to run the markets selling the meat, dairy, and egg products.

16. The supplies, fuel, and water used to manufacture all the equipment used to advertise the meat, dairy, and egg products – including advertising and marketing in other countries to increase exports and imports of feed crops, meat, dairy, and eggs.

17. The supplies, fuel, and water used to produce advertising and to advertise the meat, dairy, and egg products – including print advertising, direct mail advertising, all sorts of outdoor advertising, and hordes of radio and TV commercial time, and the continual Internet presence.

18. The supplies, fuel, and water used to keep and display the meat, dairy, and egg products in stores.

19. The supplies, fuel, and water used to manufacture and maintain all the hundreds of millions of refrigerators, freezers, stoves, and microwaves used in the stores, restaurants, hotels, homes, schools, hospitals, prisons, military bases, and on ships, and other places where meat, dairy, and egg products are kept and prepared for consumption.

20. The supplies, fuel, electricity, and water used to operate all the hundreds of millions of refrigerators, freezers, stoves, ovens, grills, and microwaves.

21. The massive amounts of water, and cleansers used to clean the farms, slaughterhouses, equipment, markets, and kitchens of all sorts where all of the meat, dairy, and eggs are produced and/or killed, sold, prepared, cooked, and eaten.

22. All the pollution left from the manufacture, marketing, packaging, and shipping of the meat, dairy, and egg products.

23. All the supplies, fuel, and water used to deal with all of the pollution.

Litter from food

When a person considers the environmental destruction caused by the meat, dairy, and egg industries, they take into account the litter from fast-food and junk food strewn among the roads, highways, and the cities, towns, and countryside, rivers, and shores. Food litter seems to be everywhere these days, especially from burger chains and junk food.

The other main types of litter are products that also lead to health problems: cigarettes and unhealthful drinks, beer and soda bottles, cans, and disposable cups discarded by the billions every day.

Bad diet fuels the medical industry

Then, there are all the resources used to run the medical industry. Much of it involves treating the results of unhealthful living, especially a diet that induces chronic and degenerative ailments. This creates and enormous amount of plastic trash and pharmaceutical pollution.

The obesity industry

The obesity industry – from liposuction to stomach stapling and intestinal bypass to "diet" pills, shots, and, increasingly, pharmaceuticals – is often the result of unhealthful living. The majority of heart surgery is the result of people not getting enough exercise and by eating unhealthful food (especially dairy, meat, eggs, and processed foods).

Killed by the unnecessary

All of this pollution, all of the animal farming, and all of the products associated with it, are not good for human, wildlife, or environmental health.

Humans do not need to drink soda or eat meat or dairy, fried foods, grilled or barbecued foods, or anything containing artificial dyes, preservatives, flavors, scents, emulsifiers, or sweeteners. The world would be a much healthier place, if humans chose healthier foods.

Fossil fuels and the animal protein diet

"This is a global business, and it's not only that we need to add to supply, but we need to reduce demand.

In the United States alone, we have about two percent of world oil reserves, five percent of the population, and yet we use about 25 percent of the world's consumption of oil."
– James J. Mulva, chairman of ConocoPhillips Co., on NBC TV's *Meet the Press*, June 18, 2006. On the same show, Shell Oil Co. President John Hofmeister explained that oil companies are holding "discussions with the White House quite frequently" with the goal of gaining greater access to U.S. federal lands (such as Nature preserves and national parks), as well as coastal waters to explore and drill for oil. Truly deplorable. In 2023, the U.S. continues to use an estimated 19 percent of the global total of oil consumption.

What industries use up the most amount of petroleum in North America?

The large majority of the food we grow on the continent – and a growing amount of creatures taken from the world's lakes and seas – is fed to farm animals. Raising, slaughtering, packaging, marketing, refrigerating, and cooking billions of farm animals every year uses tremendous amounts of fuel. Currently, it is fossil fuels.

Carbon dioxide from meat and dairy diet vs vegan diet

"The typical U.S. diet, about 28 percent of which comes from animal sources, generates the equivalent of nearly 1.5 tons more carbon dioxide per person per year than a vegan diet with the same number of calories... By comparison, the difference in annual emissions between driving a typical saloon [sedan] car and a hybrid car, which runs off a rechargeable battery and gasoline, is just over one ton."
– According to study done at the University of Chicago; It's better to green your diet than your car, *New Scientist* magazine, Dec. 2005

192

"Raising animals for food requires more than one-third of all raw materials and fossil fuels used in the United States. Producing a single hamburger patty uses enough fossil fuel to drive a small car 20 miles and enough water for 17 showers."
– PETA, VegNow.Com

If you want to help save Nature

If you want to help save Nature and protect the environment – from the sky to the bottoms of the seas – stop eating meat (including mammals, birds, and sea creatures), dairy, eggs, and processed foods.

Following a vegan diet is the single most effective way you can improve the environment, the conditions for wildlife, and the health of Earth, and prevent land degradation, rainforest and grassland destruction, climate change, ocean acidification, ecocide, and extinction.

"If anyone wants to save the planet, all they have to do is just stop eating meat. That's the single most important thing you could do. It's staggering when you think about it. Vegetarianism takes care of so many things in one shot: ecology, famine, cruelty."
– Paul McCartney, vegetarian, animal rights proponent, and music maker

The right to speak up about animal farming industry

If you are a meat eater and think eating meat, dairy, and eggs is your business, and none of mine, think again. Similarly to how *Mad Cowboy* author Howard Lyman and others have stated it, I repeat the defense: If your diet style is relying on the mass breeding of animals and killing massive numbers of sea creatures, and practices that destroy the climate, wildlife, and Nature, then it is my business, and it is the business of everyone on the planet. It affects us socially, physically, economically, and environmentally, and alters the biodiversity of life on Earth. Your diet style pollutes the air I breathe, the water I drink, and the food I eat. It has led to – and is leading to – the extinction of species. The entire animal farming industry and its feed crops and support systems collectively are the main causes of global warming.

A vegan diet free from junk food uses substantially fewer resources than a wasteful and unhealthful diet rich in carcass and dairy.

A diet rich in dairy, eggs, and animal corpse is the ecocide diet.

Plant-based diet safer for farmers, land, water, air, and wildlife

A plant-based diet is not only healthier for humans, but also for farm workers, for the land, for plant life, for the water, for the air, for the animals, for Earth, and for you.

Daniel John Carey

Greenhouse gas emissions and beef

"A kilogram of beef is responsible for more greenhouse gas emissions and other pollution than driving for 3 hours while leaving all the lights on back home. In other words, a kilogram of beef is responsible for the equivalent of the amount of CO_2 emitted by the average European car every 250 kilometers, and burns enough energy to light a 100-watt bulb for nearly 20 days."
— *New Scientist Magazine*

Hog Farm Pollution

Burning candles to tolerate the warehouse pig farm odor

"In eastern North Carolina, where pig farms outnumber people 30-1, it's not uncommon for folks living near the region's many hog operations to keep candles burning to tolerate the odor.

The landscape is too often interrupted by warehouse-sized holding pens, facilities that turn hog waste into energy that the industry calls biogas and claims is good for the environment, and industrial hog operations that store and spray billions of gallons of hog feces and urine onto nearby fields."
– Samantha Baars, *North Carolina's Hog Problem*; Southern Environmental Law Center; Jan. 2023

Natural life of pigs

When pigs live in a natural, wild environment they explore many miles per day, and sleep with other pigs in a bed of twigs, foliage, or grass. They are clean, smart, and social animals who care for their young, play, share food, watch out for each other, and enjoy the sun.

Severed pig heads processed into food and other products

"Every hour, more than 1,300 severed pork heads go sliding along the belt. Workers slice off the ears, clip the snouts, chisel the cheek meat. They scoop out the eyes, carve out the tongue, and scrape the palate meat from the roofs of mouths. Because, famously, all parts of a pig are edible ('everything but the squeal,' wisdom goes), nothing is wasted."

– Ted Genoways, *Cut and Kill*; *Mother Jones* magazine, July/Aug, 2011

Reality of pig factory farms: filth, misery, stench, noise, and pain

Today most pigs raised for human consumption are raised indoors in smelly, filthy, noisy conditions with cement floors that deform their feet. They are kept in cramped pens and fed horrible diets that are a far cry from what they would naturally eat. Pregnant pigs are kept in narrow cages restricting their movement so much they can't turn, or stretch. Often the pregnant pigs are kept with a chain or leash around their neck that is tied to the floor or pen.

Piglets are cruelly physically altered, without anesthesia

Shortly after the babies are born the piglet's teeth are cut, their tails are sliced off, and their ears are clipped off – with no anesthetic applied. Then they are placed in a pen where they will spend their entire lives with hundreds and often thousands of pigs living a similar grim fate of confinement. They are limited to eating an unhealthy diet meant to make them as fat as possible as fast as possible.

Most U.S. pigs have respiratory infections at the time of their slaughter. Several hundred thousand of them are slaughtered every day in the U.S., while more than that are being born. Behind chickens and fish, pigs are the third most commonly slaughtered animal. In 2019, about 1.3 billion pigs were slaughtered globally.

The number of factory-farmed animals

"The scale of humanity's meat consumption is enormous. 360 million tonnes of meat every year.

The number is so large I find it impossible to comprehend.

I can imagine a future in which our grandchildren look back at our time and find it difficult to believe we today are living in a world in which we kill hundreds of millions of fish, 900,000 cows, 1.4 million goats, 1.7 million sheep, 3.8 million pigs, 11.8 million ducks, and more than 200 million chickens every day."

– Max Roser, *How many animals get slaughtered every day?*; OurWorldInData.org; Sept. 2023

"More than 100 billion animals are killed for meat and other animal products every year. That's hundreds of millions of animals every day.

There is no specific definition of a 'factory farm.'

In agriculture research, they are often known as 'concentrated animal feeding operations' (CAFO).

The US Department of Agriculture has consistent criteria for CAFOs to track and quantify these farms. An animal feeding operation (AFO) is an operation where animals are kept and raised in confined situations.

It is estimated that 99% of livestock in the U.S. were factory-farmed in 2017. That was 10 billion animals. More than the global human population.

This share varied by the type of animal.

All fish raised in fish farms were considered to be factory-farmed. More than 98% of chickens, turkeys, and pigs were. Cows were a bit more likely to be raised outside in fields, with greater space and freedom. Nonetheless, 70% were still fed in concentrated feeding operations for at least 45 days a year.

Statistics for number of animals in factory farms in the U.S.:

Chickens: 8.9 billion factory-farmed. 3.6 billion not factory-farmed.

Farmed fish: 520 million factory-farmed (100%). 0 not factory farmed.

Egg-laying hens: 362 million factory farmed (98%). 7 million not factory-farmed.

Turkeys: 285 million factory-farmed (99.9%). 430,000 not factory-farmed.

Cows: 66 million factory-farmed (70%). 28 million not factory farmed.

Pigs: 71 million factory farmed (98%). 1.3 million not factory-farmed.

It is estimated that [globally] three-quarters (74%) of land livestock are factory-farmed. That means at any given time, around 23 billion animals are on these farms."

– Hannah Ritchie, *How many animals are factory farmed?*; OurWorldInData.org, Sept. 2023

The suffering of farmed pigs

"Basically, pork producers figured out some years ago that if they packed the maximum number of pigs into the minimum amount of space, if they pinned the creatures down into fit-to-size iron crates above slatted floors and carved out giant 'lagoons' to contain the manure – if they turned the 'farm,' in short, into a sunless hell of metal and concrete – it made everything so much more efficient. An obvious cost-saver, and from the industry's standpoint, that should settle the matter.

… It turns out that when you trap intelligent, 400- to 500-pound mammals in gestation crates 22 inches wide and seven feet long, when their limbs are broken from trying to turn or escape and they are covered in sores, blood, tumors, 'puss pockets,' and their own urine and excrement, they tend to act up a bit.

Indeed, the most notable thing is how the appearance of any human being causes a violent panic. A mere opening of the door brings on a horrific wave of roars, squeals and cage rattling from the sows. Another memorable sight is the 'cull pen,' wherein each and every day, the dead or dying bodies of the weak are placed, the ones who expired from the sheer, unrelenting agony of it.

– Matthew Scully, author of *Dominion: The Power of Man, the Suffering of Animals, and the Call to Mercy*; in an article published in the *Arizona Republic*; Feb. 2006

Enormous amounts of pig waste = environmental disaster

"A typical pig factory farm generates raw waste equivalent to that of a city of 12,000 people."
– PETA, VegNow.Com

"In the summer of 2022, on behalf of the Environmental Justice Community Action Network and Cape Fear River Watch, SELC challenged a one-size fits all permit issued by the North Carolina Department of Environmental Quality that would allow hog operations to use giant pits of untreated hog feces and urine to produce methane gas, or biogas, while spraying the harmful waste on surrounding fields with minimal oversight.

This outdated practice can increase water and air pollution, and continues a long history of harm to the families – disproportionately Black and Brown, Latino, and indigenous people – living nearby. "
– Samantha Baars, *North Carolina's Hog Problem*; Southern Environmental Law Center; Jan. 2023;

Landmark journalism report on pig farming: *Boss Hog Series*

Raleigh, North Carolina, *News and Observer* newspaper reporters Joby Warrick and Pat Stith, along with editor Melanie Sill, worked for several months on a series of articles exposing the environmental disasters and health risks of hog farming in factory farms in North Carolina. In April 1996 the newspaper won the Pulitzer Prize gold medal for meritorious public service.

The articles, known as the *Boss Hog Series*, told how politicians involved in hog farming helped pave the way for a billion dollar hog farming expansion where corporations are taking over the industry and elbowing out the small farmers. The politicians did so by influencing state agencies, introducing bills, helping to pass laws, and forming policies providing weaker environmental regulations and zoning protection for corporate hog farming. Many legislatures in North Carolina have a history of receiving money from the hog industry, and some are investors in hog farming.

The expansion of big business hog farming in North Carolina was done in spite of complaints and concerns voiced by long-time residents, local leaders, and environmental groups. But key to the rapid expansion is the way self-serving politicians, along with contributors to political campaigns, make millions from the hog industry.

The issues detailed in the *News and Observer* are not confined to North Carolina; they are issues being faced by farming and business communities all over the world. Corporate pig farm interests have spent millions to defeat legislators who were working to clean up the environmental hazards of factory hog farm cesspools.

Pig farms: urine, feces, flies, smells, and water, land, and air pollution

The massive amounts of pollution from the hog farms in eastern North Carolina come from housing several million hogs in large steel barns (factory farms). The combination of the hogs, barns, flies, and pollution has changed the landscape, real estate value, smell, and water quality of the region. The odors from the hog farms have invaded homes, schools, churches, and businesses. The residents are angered their communities are being spoiled by the rapidly expanding hog industry.

The hog population in North Carolina resulted in the state jumping from the seventh to the second largest hog-farming state in the U.S. Since the 1990s there are more hogs than humans in North Carolina. Each hog produces two to four times as much waste as the average human.

Origins of pigs and wild boars in America

Pigs are not native to the Americas. Pig fossils have been found in Europe and Asia. It is believed they were first domesticated in the Near East around 8,500 B.C. and in China about 4,900 B.C. They were brought to America by the Europeans and released into – or escaped into – the wild in the 1500s.

In North Carolina, by 2020, the wild boar population was estimated to be 100,000. Humans have made pigs a globally invasive species.

More pigs than humans

In 2022, the farmed pig population in North Carolina was about 8.1 million. The state's human population in 2021 was 10.55 million.

With 23.4 million pigs in 2022, Iowa was the number one pork-producing state. The human population of Iowa is 3.193 million.

Hog farms in North Carolina produce more than ten million tons of waste every year. With millions of hogs in such a small area producing so much waste, North Carolina hog farms are turning out more raw sewage than both New York City *and* the suburbs surrounding it.

The pollution produced by the hog farms in North Carolina is overwhelming the land and the people who live there. The situation has been able to magnify because North Carolina has weaker and imposes fewer environmental regulations for hog farms than any major hog-producing state.

Trillions of pounds of farmed animal waste

"Animals raised for human consumption in the U.S. generate 2.2 trillion pounds of waste each year."
– A Voice for Animals, VoiceForAnimals.org; 2002

The putrid waste from the hog farms is stored in thousands of cesspools the hog industry likes to charmingly refer to as "lagoons." Those in the hog industry say the hog waste safely decomposes in the cesspools before it is sprayed onto cropland. The problem is the land cannot absorb so much waste. The waste stored in the cesspools also leaks into ground water.

"There are a lot of those lagoons in North Carolina, a state where the pork industry brings in billions of dollars. The state has said at least 50 lagoons overflowers in the wake of Hurricane Florence, and many more have the potential to spill over as well."
– Wynne Davis, *Overflowing Hog Lagoons Raise Environmental Concerns in North Carolina*; NPR Weekend Edition Saturday, NPR.org; Sept. 2018

Hog waste lagoons poison water, kill wildlife, and ruin farmland

Hurricanes are a problem for hog waste lagoons, causing the lagoons to spill millions of gallons of toxic waste into rivers, creeks, wetlands, and marshes. The waste rich with bacteria, nitrates, ammonia, feces, urine, blood, and afterbirth, also gets into flooded homes, schools, businesses, and other structures. Vegetable and fruit farms also are poisoned during these floods, ruining crops, and perhaps causing the land to be unusable for years. These situations have been increasing.

Animal farming contributing to global warming and super storms

As we are seeing, as the climate heats up, the clouds hold more water, and the heavy rainstorms pour more inches of rain, which is increasing flooding everywhere from cities to farm communities. I am writing this paragraph in 2023, days after a massive rainstorm flooded parts of New York City in ways never previously experienced. Expect more of that in coming years.

Hog farms exempt from environmental rules, and cause illness

"When it pertains to environmental degradation, large hog farms are polluting the air and water with hazardous chemicals, including carbon monoxide, methane, antibiotic residues, and pathogenic bacteria. Some research suggests there is an increased occurrence of respiratory illnesses, asthma, and even cancer among people living near hog farms.

Even though it may not affect your community directly, hog farms are often exempted from environmental standards linked to increases in environmental disruption, potential for disease outbreaks, and, of course, environmental injustice.

It is estimated that pigs produce about 11 pounds of manure daily. This waste, just like human waste, has to be stored. However, while humans have an intricate sewage system, pigs do not.

Pig manure is stored in on-farm lagoons, or small ponds. These lagoons overflow or are released into crop fields as fertilizer that goes on to emit toxic fumes. This method is known as the 'lagoon spraying method.' This spraying method is what often results in the reported air pollution. Also, lagoons can overflow and cause bacteria to contaminate water."

– Will Burnett, *NC Talks: Hof farms' effect on local communities and the environment*; *Carolina Public Press*; Sept. 2023

Hog farms dump waste and poison drinking water

The *60 Minutes* TV show reported in December 1996 that over 30 percent of water wells near hog farms were contaminated by hog waste. The cesspools also overflow into streams and rivers during heavy rains. Millions of gallons of hog farm sludge, hog feces, hog afterbirths and blood, as well as cropland fertilizer have leaked into surrounding waterways where it has been killing aquatic life and causing algae overgrowth that chokes waterways. In one incident an estimated 25 million tons of hog waste flowed out of an eight-acre cesspool when a levee broke. Many farms have been caught dumping hog waste directly into surrounding streams, rivers, and lakes. The pollution has killed millions of fish and many other types of waterlife, and caused rivers and lakes to be closed off to swimming and sports.

This is all done so humans can eat bits of pig corpse, which increases human cardiovascular disease, cancers, and other health issues.

One hog farm cesspool spill killed millions of fish

In 1995 a 120,000-square-foot hog cesspool released over 25 million gallons of waste into the headwaters of North Carolina's New River. It took months to reach the ocean, and killed millions of fish and unknown numbers of animals unfortunate enough to be in the contaminated river.

Blood, urine, feces, puss, chemicals, and drugs

The leak above was one of many from cesspool lagoons in North Carolina. Some go unreported, because the authorities are not informed. They don't have the staff to supervise the massive numbers of factory farms in the state, or to monitor the environment surrounding each factory farm – which are often near the slaughterhouses and rendering facilities that also cause massive pollution. That is, releasing massive amounts of blood, urine, feces, and puss into the environment – and also bacteria, viruses, and the chemicals and pharmaceutical drugs (including Ractopamine [research that drug and its side effects on human health]) used on the farmed animals.

As you read this about factory farm pollution, magnify the environmental impact when considering other branches of the animal farming industry. This includes the slaughterhouses, rendering facilities, and all the fuel and plastic, electricity, and fossil fuels used to process, package, refrigerate, transport, market, and cook the meat.

Pig farms pollute wetlands, creeks, rivers, lakes, and marshes

"A Jones County farm that spilled an estimated one million gallons of hog feces and urine into a tributary of the Trent River shortly before Christmas had been cited twice in the past year by state regulators because its lagoon was too full.

Because of the location of DC Mills Farms, even a minor accident can cause major problems. The farm lies on a peninsula of land within a flood zone that includes a swamp. Tuckahoe Creek runs through the farm and adjacent to the hog lagoon, before joining the Trent River, two and a half miles away."
– Lisa Sorg, *Hog farm that spilled 1 million gallons of feces, urine into waterways had been warned of lagoon problems*; NC Newsline; Jan. 2021

The above article details numerous other leaks from cesspool lagoons on various North Carolina factory pig farms.

Of pig pollution, human diseases, and deaths

"North Carolina is the second largest producer of pork in the country, and Sampson, Duplin, and Bladen counties house more than

40% of the state's hog farms. According to census data, between a quarter to a third of the population in all three of these counties are Black; around a quarter of Duplin and Sampson residents are Latinx. About 42% of residents in neighboring Robeson County – the fourth largest hog producing county – are Native American.

Hog waste – feces, urine, blood, and pus [and afterbirth] – drips through slots in the floors and it empties into open pit lagoons, where it is mixed with water. This wastewater often sits uncovered, emitting noxious odors, and gasses including methane, or is pumped out and sprayed as fertilizer on crop fields. There are approximately 4,000 hog farm lagoons in North Carolina, and the sludge – unusable waste left after the liquid has been sprayed – has almost filled many of these lagoons to capacity.

The pollution from concentrated animal feeding operations disproportionately impacts rural Black, Latinx, and Indigenous communities in North Carolina, including the Hiliwa, Saponi, Coharie, and Lumbee tribal nations. Studies have shown they're more likely to live within three miles of large swine operations. A 2021 report also found that in Duplin County, 89 premature deaths per year can be attributed to emissions from hog operations."

– Cameron Oglesby, 'This plan is a lie': Biogas on hog farms could do more harm than good: North Carolina residents, researchers, and farmers say the rapidly growing industry distracts from a massive hog waste problem – and the public risks that it causes; Energy News Network; March 2022

Thousands of drowned, bloated pigs in rivers to the ocean

In 1999 Hurricane Floyd caused so much flooding in Eastern North Carolina that well over 120 million gallons of hog waste made its way into the rivers – poisoning them, and out to the sea. It carried with it thousands of drowned, bloated pigs, and killed unknown millions of fish. There were too many dead fish for birds to eat, and too many pigs in the ocean for sharks to feast on.

Some form of the above story keeps repeating, each time a hurricane or massive rainstorm goes through the region. Pigs, cesspools, floods, environmental devastation, dead wildlife, poisoned water. Repeat.

Nitrates from hog cesspools cause methemoglobinemia

Nitrate-nitrogen from the hog cesspools continues to leak into ground water. This is a major problem because the chemical causes methemoglobinemia, a disease that hampers the ability of the blood to absorb oxygen. It can be particularly lethal to infants who drink contaminated water. All so people can eat pig carcasses.

"Nitrogen-rich animal waste suspected"

"This year the Environmental Protection Agency found that 21 percent of the groundwater sampled in Washington's agricultural Yakima Valley contained unsafe levels of nitrates, leaving it unfit for residents to use. A final report is due out in coming weeks, but nitrogen-rich animal waste is a suspected contributor."

– Monica Eng, *The costs of cheap meat: Critics of factory farms say we pay a high price for low-cost food; Chicago Tribune*, Sept. 2010

It should be no surprise that animal farming is the cause of nitrates poisoning groundwater used for everything the people need it for. It sure isn't from butterflies, birds, trees, or other forms of wildlife. It's human-caused via animal farming practices.

Nitrate pollution is a problem wherever there are factory hog farms

The problems with nitrate pollution exist wherever there are hog farms. This is especially so with the large factory warehouse "barns" confining thousands of pigs. They spend their lives in noisy misery on concrete and metal flooring, while eating horrible diets to fatten them. The cesspools containing hog feces, urine, puss, blood, afterbirth, viruses, and drugs, end up doing awful things to the environment.

In Nebraska there are tens of thousands of factory hog barns in an area continually testing high in nitrate pollution.

Nebraska Governor Jim Pillen's family hog farm mess

Many of the hog facilities in Nebraska are owned by Governor Jim Pillen, who grew up there, and helped to transform his family farm into an empire producing hog meat sold on the global market. The nitrate pollution has saturated the ground so deeply that wells have become too contaminated to use. One cause is the nitrates sprayed on feed crop fields – specifically corn. Another cause is pollution from the hog facilities. The hog facilities owned by the Pillen family are some of the worst polluters. They are required to have monitoring wells to test the ground water. Not all factory hog farms comply with the rule. Some of the ground water is so contaminated with nitrates that it can be used as fertilizer.

With factory hog farms going global, so are the problems they cause

These issues of hog facilities in the U.S. making groundwater toxic are in a country that supposedly has environmental protections. Who knows what is happening with the growing number of factory hog farms in other countries.

Just as with other industries, if it gets cheaper to produce a product outside of the U.S., that is likely where more and more hog farming will go – including because other countries could be more lax on hog facility pollution. The ammonia, nitrates, and leaking hog waste lagoons poisoning their water, damaging the environment, and killing wildlife could go unchecked.

Cancer from nitrate pollution

"High nitrate has been linked to a variety of health conditions, including cancer in children.

Nebraska has the highest pediatric cancer rate west of Pennsylvania. Counties with higher nitrate levels often have higher rates of pediatric cancer.

Andrew Greisen, Platte Center's water operator, says the area surrounding the town has seen several cancer cases this year.

'Prostate cancer, breast cancer, and brain cancer, just everything,' he said. 'I just think it's got to be the food we're eating, or the water we're drinking.

…

In 2006, a U.S. Fish and Wildlife employee reported that workers at a Hastings-area hog farm were pumping hog waste onto a nearby federal wetlands area. In a separate indecent, state regulators alleged, farm employees used PVC pipe to drain a storage pit into a freshwater channel."

– Yanqi Xu, *Pillen's Water: High nitrate detected on hog farms owned by Nebraska's governor*; *Flatwater Free Press*, NebraskaPublicMedia.org, Sept. 2023

Hog farms: bacteria, viruses, parasites, and human disease

Other ingredients of hog waste that may threaten human and animal health include bacteria, viruses, and parasites. Pfiesteria piscicidia is one of the microbes. It results in massive fish kills. When humans are exposed to this microbe they can experience skin sores, nausea, vomiting, headache, blurred vision, breathing difficulties, liver and kidney problems, memory loss, and cognitive impairment.

The putrid smell from the hog farms can saturate everything in the community, including well water, food, laundry, furniture, and the structures of the homes.

Dumping dead hogs

In addition to the hog waste polluting the air, water, and land of North Carolina, the people regularly find dead hogs dumped in the countryside. The corporation may own the pigs, but the farmers who are under contract with the corporation are supposed to take care of the pigs that die, and the mass quantities of hog waste. When farmers under contract raise thousands of pigs they can quickly get overwhelmed with the waste and carcasses.

What is a growing problem in North Carolina is not limited to that state. The corporate players in the hog industry have expanded in several states as well as to other countries, including Eastern Europe. The hog pollution experienced in North Carolina and Iowa is a growing problem in Poland.

What's in those cesspool "lagoons"?

Some factory hog farm facilities are designed to hold several hundred thousand hogs with adjoining cesspools holding millions of gallons of urine mixed with feces containing all the pharmaceutical drugs the animals were given, all of the toxic sprays they were treated with, all of the farming chemicals used on the foods the pigs ate, and a wide variety of bacteria. In addition, as mentioned earlier, the lagoons contain the afterbirth of hogs, an also stillborn pigs and piglets that died soon after birth.

Smithfield Farms slaughters millions of pigs

One hog factory in Utah has about 500,000 pigs. Since hogs produce about three times as much excrement as humans, those 500,000 hogs produce more waste than 1.5 million people. It is a subsidiary of Smithfield Foods, a company responsible for the slaughter of about 27 million hogs in 2005.

An article that appeared in *Rolling Stone* compared this Smithfield Foods slaughter in body weight to killing all of the human population of America's 32 largest cities (Boss Hog, by Jeff Tietz, *Rolling Stone*, Dec. 2006).

Factory farm employees falling in and dying in pig cesspools

The same article estimated the excrement from Smithfield Foods hog farms for one year would fill four Yankee Stadiums. The article tells of factory hog farm workers who have died in these cesspools of hog waste, including five members of one family.

From the lips of this guy, rich from pig torture

"The animal-rights people want to impose a vegetarian's society on the U.S. Most vegetarians I know are neurotic."
– Joseph Luter III, Chairman of Smithfield Foods, the world's largest pig farming company. Producing billions of pounds of hog meat every year. from Boss Hog, by

Jeff Tietz, *The EPA has cited Smithfield Foods for thousands of violations of the Clean Water Act; Rolling Stone*, Dec. 2006.

If you eat ham, bacon, pork chops, or other pork, you are supporting this horrible hog industry, and the environmental destruction it causes.

Some of those making the money from factory farming and the meat industry claim they are doing a service for humanity by producing meat, dairy, and eggs. As if all of the problems the animal farming industry creates are validated. Humans don't need to eat animal protein.

"When nonvegetarians say that 'human problems come first' I cannot help wondering what exactly it is they are doing for human beings that compels them to continue to support the wasteful, ruthless exploitation of farm animals."
– Peter Singer, *Animal Liberation*, 1990

Using the need for protein as a need to continue killing animals

Some people claim humans "need animal protein." The human body makes protein from – and proteins are made of – amino acids. The amino acids the human needs for making protein are found in fruits, vegetables, sprouts, nuts, seeds, and seaweeds. To think that you need to eat animals that eat plants to get the protein from them that you can get by eating a variety of plants is irrational. Considering all of the destruction, misery, pain, illness, climate change, extinction, and other issues resulting from the animal farming industry, the reliance on carcasses, cow milk, and what are essentially chicken periods for protein is tragically ironic.

Horses, giraffes, gorillas, elephants, antelope, deer, and other plant-eating wild animals seem to be doing quite fine without eating corpse, milk, and eggs to get protein to build their strong muscles.

My diet has been vegan for over 34 years. I run in the sand for miles, bike for miles, do yoga, and other exercise. Doctors say I have the heart of a teenager. (See my book: *Plant-based Regenerative Nutrition*.)

"One of the topics that are still used in the West to justify the annual massacre of hundreds of millions of chickens, lambs, pigs and cattle for sustenance is that you need protein. And the elephants? Where do elephants get protein?"
– Tiziano Terzani

"People often say that humans have always eaten animals, as if this is a justification for continuing the practice. According to this logic, we should not try to prevent people from murdering other people, since this has also been done since the earliest of times."
– Isaac Bashevis Singer

"Our treatment of animals will someday be considered barbarous. There cannot be perfect civilization until man realizes that the rights of every living creature are as sacred as his own."
– Dr. David Starr Jordan

Meat allergy: alpha-gal syndrome and the lone-star tick bite

For some people, their meat and dairy consumption will end in ways out of their control. This is because of a meat allergy some people are getting. It was identified in the U.S. in 2008.

The allergy called alpha-gal syndrome being spread by the saliva from a bite by the Amblyomma Americanum "lone star" tick. The back of the tick features a white spot.

The global warming temperatures are increasing the population and range of the lone-star tick.

Humans bitten by the lone-star tick who then eat meat can experience diarrhea, hives, shortness of breath, tightness in the throat, and belly cramps.

Severe reactions to alpha-gal syndrome: potentially fatal

Under severe reactions, a person can experience anaphylaxis – with the body releasing chemicals that can cause confusion, itchy skin, rashes, swelling of the hands and feet, swollen eyes, lips, tongue, and throat, nausea and vomiting, and swollen breathing passageways, wheezing, lightheadedness, fainting, unconsciousness, and shock. It can result in death.

The meat and meat products people with the alpha-gal "meat allergy" need to avoid include beef, dairy products, lamb, pork, rabbit, venison, and even gelatin, breads and other products containing dairy – and pharmaceutical drugs and nutrition supplements containing gelatin or in gelatin capsules.

Potentially life-threatening allergic condition

According to the U.S. Centers for Disease Control and Prevention, from 2010 to 2021, over 110,000 cases of alpha-gal syndrome have been reported. The condition is increasing. Some people who experience light symptoms may not detect something is wrong, or, because meat and dairy can take hours to digest, they might not relate their light symptoms – such as puffy eyes and nasal congestion – to having had ingested the animal protein.

"Alpha-gal syndrome (AGS) is a serious, potentially life-threatening allergic condition. AGS is also called alpha-gal allergy, red meat allergy, or tick bite meat allergy. AGS is not caused by an infection. AGS symptoms can occur after people eat red meat, or are exposed to products containing alpha-gal.

It is not known how many cases of AGS exist in the United States. Additional data and research are needed to understand how many people are affected by this condition.

Alpha-gal is a sugar molecule found in most mammals. It is not found in fish, reptiles, birds, or people. It can be found in meat (pork, beef, rabbit, lamb, venison, etc.), and products made from mammals (including gelatin, cow's milk, and milk products).

AGS is diagnosed by an allergist, or other healthcare provider, through a detailed patient history, physical examination, and a blood test that looks for specific antibodies (proteins made by your immune system) to alpha-gal."

– U.S. Centers for Disease Control and Prevention; 2023

Australia and their tick-bite animal protein allergy

A similar meat allergy has been spreading in Australia, and is the result of a different tick, the Ixodes holocyclus.

Gelatin

Because gelatin is usually made from animals, people with the allergy should look for vegetarian or vegan gel caps when purchasing vitamins and other supplements. And look for vegan supplements. They should also avoid products containing the dairy products casein and whey.

War on Wildlife:
Killing Millions of Wild Animals
to Protect Farmed Animals

Globally, 92.2 billion animals slaughtered

"The most recent data about land animals bred, kept, and slaughtered for consumption has revealed a figure higher than ever before: An estimated 92.2 billion land animals are slaughtered annually in the global food system, according to the Food and Agriculture Organization."

– Kitty Block, More animals than ever before – 92.2 billion – are used and killed each year for food; Blog.HumaneSociety.org, June 2023

AnimalClock.org counts animal slaughter

The site AnimalClock.org provides a continual count of how many animals are killed for food in the U.S. per year. As I'm writing this paragraph at the end of December 2023, the number has surpassed 55 billion. This includes both mammals, birds, fish, and shellfish.

Cattle, pigs, turkeys, ducks, chickens, goats, lambs, fish, and other animals aren't the only animals being killed by the animal farming industry.

Wildlife killed by meat, dairy, egg, and feed crop industries

Massive numbers of wild animals have been – and are being – killed by meat farmers, by corporate farms, and by governments working to protect farmed animals and farmed animal feed crops.

Bison, cougars, coyotes, foxes, wolves, and other wildlife killed

"Every year, tens of thousands of bison, coyotes, wolves, and other wildlife are maimed, shot, poisoned, and even burned alive because the meat industry claims these animals interfere with raising animals for food. This war on wildlife is carried out with the full support of state and federal agencies, which fund so-called predator control programs."
– FarmSanctuary.Org

Meat industry: the killing machine

"Animal agriculture directly kills annually nearly 50 billion animals worldwide, after subjecting them to the cruelties of factory farming. It also kills uncounted numbers of wildlife on land and in the seas."
– Citizens for Healthy Options in Children's Education, CHOICE.USA

Livestock: number one source of water pollution

"But it [cattle ranching] is anything but benign. It is the number one source of water pollution in the West. It's the number one source of soil erosion in the West. It's the number one cause of species endangerment in the West. It's the reason we don't have wolves throughout the West. It's one of the major reasons that more than four-fifths of all native fish west of the Continental Divide are endangered or threatened."
– Tim Lenerich, *Dispelling the Cowboy Myth*; George Wuerthner of Eugene, Oregon, is one of the most outspoken leaders against public-lands ranching. EarthSave.Org; 2004

U.S. Government: cattle rancher subsidies and killing wildlife

"In reality, ranchers are the most pervasively destructive force on our public land, with logging as a distant second. Via outlandish subsidies, you, I, and Uncle Sam support the cattle industry with drought and fire relief, fencing, water tanks, windmills, and bargain-basement grazing fees. Our government kills hundreds of thousands of wild creatures each year to protect ranchers' herds against predators such as wolves, mountain lions, and coyotes.

In return we get erosion, endangered species, habitat destruction, flash floods, exotic weeds, desertification, and some of the most degraded landscape on Earth."
– Tim Lenerich, *Dispelling the Cowboy Myth*, Earthsave.Org; 2004

"Nearly 20 million taxpayer dollars fund the trapping, poisoning, and shooting of native predators deemed a threat to agriculture by the USDA Wildlife Services Agency, which each year kills approximately

211

100,000 coyotes, bobcats, foxes, bears, wolves, and other predators. In 2001 the program also killed 1.6 million other 'nuisance' animals.

Almost two-thirds of all large mammal species are threatened or endangered in the lower 48 states. Less than 10 percent of all endangered and threatened species in the U.S. is improving.

About 20 percent of all endangered and threatened species are harmed by grazing."
– A Voice for Animals, VoiceForAnimals.org; 2002

Pollution takes an enormous toll on wildlife populations. Other reasons populations of native animals have plunged include:
1) Wild animals have been killed to protect livestock.
2) Wild animals have been killed to protect feed crops.
3) Habitat land cleared for feed crops, and to build roads and airports, and for golf courses, mining, trash dumps, and urban sprawl.
4) Forests are cut for timber, paper, packaging, and urban sprawl.
5) Insect populations are dying from chemicals used on crops, sporting fields, and landscaping. This is also plunging bird populations. Humans can't survive without insects or birds.

"Wildlife Services," the U.S. Government wildlife killing team

"In response to ranchers' complaints of coyotes attacking cattle in southern Arizona, the federal government took to the air this past January, killing 200 coyotes. The hunt was conducted by Wildlife Services, a division of the U.S. Department of Agriculture, and took place on both public and private land."
– Earth First! Journal, EarthFirstJournal.Org; May-June 2006

To farm animals for meat, native animals need to be eliminated

To make it possible for cattle and other livestock to graze in open fields, or even live in fenced-in areas, there has to be a safe area created for them. This means native animals have to be eliminated. In North America, these animals include wolves, coyotes, foxes, lynx, mountain lions, bobcats, jaguars, bears, elk, bighorn sheep, deer, porcupine, beaver, badgers, skunks, possums, prairie dogs, and antelope.

U.S. "Animal Damage Control" formed to exterminate wolves

"In 1914, Congress first appropriated money for the U.S. Biological Survey (now known as Animal Damage Control) to exterminate wolves from the face of the continent. Though the agency failed to eradicate the species entirely, by 1945 it had killed every wolf in Colorado."
– Colorado Wolf Tracks; 1996

The list of animals killed to protect feed grains and farmed animals

Throughout North America, early homesteaders as well as modern ranchers and other livestock farmers have killed off much of the predator animals, including mountain lions, bobcats, foxes, wolves, bears, and jaguars. This has increased the populations of many animals, including deer, squirrels, raccoons, mice, rabbits, and groundhogs. Many of those smaller animals also are problematic to ranchers. Ground animals build dens and tunnels that can cause cattle to injure their legs. To get rid of unwanted wild animals, including ravens and birds that eat grain, or may otherwise damage crops, farmers and ranchers kill millions of animals every year by poisoning them, burning them, flooding their dens, shooting them from airplanes and helicopters, and electrocuting them. Deer populations are culled by allowing for "hunting season."

Eliminating a species from its terrain changes the land, all the way down to the microorganisms, and how and which plants grow there. It can also bring in other animals who otherwise would not be living there. Consider this from Peter Wohlleben's book *The Hidden Life of Trees*:

What eliminating wildlife does

"Wolves disappeared from Yellowstone, the world's first national park, in the 1920s. When they left, the entire ecosystem changed. Elk herds in the park increased their numbers and began to make quite a meal of the aspens, willows, and cottonwoods that lined the streams. Vegetation declined, and animals that depended on the trees left. The wolves were absent for seventy years. When they returned, the elks' languorous browsing days were over. As the wolf packs kept the herds on the move, browsing diminished, and the trees sprang back. The roots of cottonwoods and willows once again stabilized stream banks and slowed the flow of water. This, in turn, created space for animals such as beavers to return. These industrious builders could now find the materials they needed to construct their lodges and raise their families. The animals that depended on the riparian meadows came back, as well. The wolves turned out to be better stewards of the land than people, creating conditions that allowed the trees to grow and exert their influence on the landscape."
– Peter Wohlleben, in his book *The Hidden Life of Trees*

Animals that increased in numbers after wolves in Yellowstone culled the overgrazing deer and elk, include bear, rabbits, mice, hawks, weasels, foxes, beaver, ravens, bald eagles, otters, muskrats, ducks, songbirds, amphibians, and reptiles. The animals spread nutrients and seeds. The land recovered. (Search: Trophic cascades.)

213

Animal agriculture harms land, air, and wildlife

The cattle industry and its use of millions upon millions of acres of land on every continent has played a role in the degeneration of wildlands, creeks, streams, rivers, ponds, lakes, marshes, and oceans. The cattle degrade the land, trample riverbeds and other waterways, damage the ability of native grasses, bushes, trees, and other plants to grow, which reduces native wildlife, and that impacts the water. This includes the acidity of the water, as plants on the land influence the chemistry of the water that falls there, and the water retention of the soil, and where the rainwater travels. That impacts the marine life, from the marine birds to the amphibians, shelled beings, and fish.

A combination of habitat loss – including of damaged, disappearing, or polluted streams, rivers, ponds, marshes, and lakes, and heating landscapes – has greatly damaged the populations of salamanders and other amphibians. Much of this can be linked to the livestock industry, cattle grazing, clearing land and altering landscapes to grow millions upon millions of acres of feed crops for farmed animals, the use of chemicals on those crops, the water and land pollution caused by the cattle, pig, chicken, and turkey farming industries, and the elimination of native plants and animals caused by all of it.

Amphibian populations globally plunging

"The updated Red List Index shows that the status of amphibians is deteriorating globally, particularly for salamanders and in the Neotropics. Disease and habitat loss drove 91% of status deterioration between 1980 and 2004. Ongoing and projected climate change effects are now of increasing concern, driving 39% of status deterioration since 2004, followed by habitat loss (37%).

The landmark 2004 Global Amphibian Assessment was published on the IUCN Red List, demonstrating that amphibians were the most threatened class of vertebrates worldwide."
– *Ongoing declines for the world's amphibians in the face of emerging threats;*
Nature.com; Oct. 2023

"Agricultural expansion" is given as the chief cause (77%) of amphibian population decrease anywhere the feed crop and livestock industries exist. The industries cause habitat damage and loss where amphibians would normally exist. Climate change plays the largest role in the amphibian loss where the cattle and feed grain industries don't have a presence. But climate change is largely caused by the meat, dairy, and feed grain industries. What those industries do is destroy the planet, and satisfying stockholders. It is, as George Tsakraklides calls it, necrocapitalism. They join the fossil fuels industries in that feat.

"A civilization which cannot see itself beyond growth is a defeated drug addict waiting for their death."
– George Tsakraklides, biologist and author of *In The Grip of Necrocapitalism: The Making and Breaking of A Psychonomy*

How U.S. government kills wildlife to protect feed crops and meat

Meanwhile, the meat, dairy, and feed crop industries continue trampling and degrading the land and water, impacting plant and animals life, and water quality, and determining which wild animals the industries allow to remain, and which wild animals they kill.

In addition to shooting them, eliminating the native animals is done with poisons, with steel jaw leghold traps, and by damaging their food supplies. The government offices involved in these activities include the Bureau of Land Management (BLM) and the Animal Damage Control (ADC) program of the U.S. Department of Agriculture.

List of wildlife killed by U.S. using poisons, guns, traps, floods, and fire

"The U.S. Department of Agriculture's Wildlife Services reported killing 404,538 native animals in 2021, according to new data released by the program today. The federal wildlife-killing program targets wolves, coyotes, birds, and other wild animals, primarily to benefit the agricultural industry in states like Texas, Colorado, and Idaho.

According to the report, the multimillion-dollar program last year killed 324 gray wolves, 64,131 coyotes, 433 black bears, 200 mountain lions, 605 bobcats, 3,014 foxes, 24,687 beavers, and 714 river otters. These figures almost certainly understate the actual number of animals killed, as program insiders have revealed that Wildlife Services kills many more animals than it reports.

'It's stomach-turning to see this barbaric federal program wiping out hundreds of thousands of native animals,' said Collette Adkins, carnivore conservation director at the Center for Biological Diversity. 'Killing carnivores like wolves and coyotes to supposedly benefit the livestock industry just leads to more conflicts and more killing. This is a truly vicious cycle, and we'll continue to demand change from Wildlife Services.

The reported number of native animals killed in 2021 was similar to the 433,192 killed in 2020. These numbers reflect a steep decline compared to 2019, when approximately 1.3 million native animals were killed. The red-winged blackbird is an example of a species with fewer individuals intentionally killed by Wildlife Services, with 15,096 killed in 2021, compared to 364,734 n 2019.

According to the new data, the wildlife-killing program unintentionally killed more than 2,746 animals in 2021, including bears, bobcats, mountain lions, foxes, muskrats, otters, deer, turtles, and dogs. Its killing of nontarget birds included wood ducks, tree swallows, herons, and turkeys. Such data reveals methods used by federal agents.

Wildlife Services poisoned 7,573 animals using M-44 cyanide bombs in 2020. Of these deaths, 314 were unintentional.

This month marks the fifth anniversary of an Idaho teen nearly being fatally poisoned by an M-44. The incident received worldwide media coverage and spurred federal and state efforts to ban devices.

'It's inexcusable that Wildlife Services continues to target rare and ecologically important animals like wolves and grizzly bears, forcing them to suffer and die in cruel traps and snares,' Adkins said. 'Taxpayer-funded wildlife slaughter needs to stop, and be replaced with a program that provides for nonlethal tools to effectively prevent most conflicts with wildlife.'

In the last few years, litigation and community opposition curtailed Wildlife Services operations in numerous states, including California, Idaho, Minnesota, and Washington, as well as localities such as Humboldt County (California), and Minneapolis."

– *400,000 Native Animals Killed by Federal Program Last Year*; Center for Biological Diversity, March, 2022

Meant to "alleviate and mitigate wildlife damage"

"The Wildlife Damage Management programs through South Dakota Game, Fish, and Parks will go through an external review to evaluate the efficacy of the programs.

Two programs, the Damage Management and Animal Damage Control, make up the Wildlife Damage Management programs meant to alleviate and mitigate wildlife damage to landowners and producers. [Ed: Landowners: cattle ranchers, feed crop farmers. Producers: of meat and feed crops.]

The Wildlife Damage Management program has been in effect since 1998 and deals mostly with game – elk, geese, turkeys – that damage private property. The Animal Damage Control program allowed GFP (game fish and parks) staff to assist landowners with wildlife control and nuisance animals."

– Makenzie Huber, *Governor requests external review for Wildlife Damage Management programs*; Argus Leader; May, 2019

"Livestock predators"

In 2019, the California Cattleman's Association provided a list of "livestock predators." They give a phone number to call USDA Wildlife Services to report when a farmed animals is killed by a predator. Their list includes:

1) Coyotes.
2) Mountain lions.
3) Domestic dogs (including wolf-dog hybrids).
4) Bears: Black and brown bears.
5) Ravens and crows. Listed because they might attack livestock, especially calves, injuring or killing them.
6) Wolves.

Wide number of wild animals "problematic" to farmed animals

The University of California Agriculture and Natural Resources Department lists other wild animals problematic to farmed animals as foxes, bobcats, raptors (golden and bald eagles), and scavenger birds (crows and magpies).

Depending on the region of the world where livestock is being raised, including cows, pigs, lambs, goats, birds, insects, and other wildlife, there are wild animals killed. Everything from alligators to ground squirrels to badgers are killed by farmers and by government agricultural departments so animals can be raised for meat, dairy, and eggs. Additionally, billions of insects, including butterflies, bees, praying mantises, and other insects are killed using chemicals to protect feed crops grown for livestock. The international meat industry is a killing machine, on every level.

The chain of destruction and degradation

Here are humans in America, government employees, meat famers, and feed crop farmers killing numerous types of wild animals, over and over, in horrible ways. It is largely done to protect and feed an animal not native to the continent: cattle. An animal brought over by the Europeans, and which have been so massively bred and fed here for centuries that having them here has been and is damaging the soil, rivers, lakes, oceans, air, climate, plants, and wild animal life. Because eating beef and dairy increases cardiovascular disease, kidney disease, diabetes, arthritis, osteoporosis, Alzheimer's, asthma, macular degeneration, and other human health maladies, the cows are ravaging human health – which feeds into the massively polluting medical and pharmaceutical industries – and all the plastic and drug pollution that leads to.

Killing California wildlife

In California, the U.S. Department of Agriculture agreed in 2020 to reduce the killing of wildlife in 10 California counties, including through pesticides, bullets, aerial gunning (shooting animals from helicopters and airplanes), and traps. This was the result of a lawsuit filed by the Earth Island Institute, the Animal Legal Defense Fund, and the Center for Biological Diversity.

An agreement in 2017 also put restrictions on killing wildlife in 16 other counties in California. The state has 58 counties.

"This victory will save hundreds of animals that would have needlessly suffered and died in traps set by Wildlife Services over the next several years."
– Collette Adkins, attorney for the Center for Biological Diversity; April 2020

"Targeted animal killings" to protect ag and livestock

"The USDA's multimillion-dollar Wildlife Services program uses several techniques – including targeted animals killings – to protect agriculture, livestock, natural resources, and properties like airports from wild animals

Three wildlife groups sued the USDA in Augusts 2019, claiming its reliance on an outdated environmental assessment from 1997 for its predatory killing program in 10 counties violates the National Environmental Policy Act.

Wildlife Services killed 26,441 native animals in California in 2018, including 3,826 coyotes, 859 beavers, 170 foxes, 83 mountain lions, 105 black bears, and 5,675 birds, according to USDA data cited in the lawsuit."
– Nicolas Iovino, *USDA Agrees to Limit Wildlife Kill Program in 10 California Counties; Courthouse News Service*, April 2020

Lead bullets and DRC-1339 poisoning wildlife, land, and water

Part of the agreement was to stop the use of lead bullets to kill wildlife, unless the carcasses can be retrieved to prevent lead from entering the environment, poisoning wildlife, and polluting land and water.

Another part of the agreement was to halt the use of the poison DRC-1339 used to kill birds, including the tricolored blackbirds, which are endangered.

All of this killing, killing, killing, and more killing. All so humans can mass breed certain animals, to kill and eat them. It's so warped, and wrong, and, ironically, damaging humanity's ability to survive.

Killing beavers

The killing of beavers has damaged the environment as the natural behavior of beavers to build dams creates areas where other wildlife thrive, including frogs, certain birds, flycatchers and other insects, and fish that all play roles in the circle of life. Nobody should be killing beavers. A July 2019 agreement with the USDA was to stop the trapping and/or killing of beavers along 11,000 miles of rivers and 4 million acres of land in the state of California.

Grip traps

Grip traps are commonly used in many countries to trap and kill wildlife considered threats to farmed animals. Being caught in a grip trap is a terrible, painful, torturous, agonizing way to die.

Killing a Noah's Ark of species

"The latest annual toll of Wildlife Services, a department within the U.S. Department of Agriculture, has further stoked the fury of conservation groups that have decried the killings as cruel and pointless. Wildlife Services maintains the slaughter is necessary to protect agricultural output, threatened species, and human health.

The 2021 toll shows the killings span a Noah's Ark of species, including alligators, armadillos, doves, owls, otters, porcupines, snakes, and turtles. European starlings alone accounted for more than one million of the animals killed. A single moose was shot, along with a solitary antelope and, accidentally, a bald eagle.

Last year's death toll was, in fact, fairly low by the standards of recent years. In both 2008 and 2010, Wildlife Services killed 5 million animals, and as recently as 2019 it killed around 1.3 million native animals, a total much higher than last year."

– Oliver Milman, 'A barbaric federal program': U.S. killed 1.75 million animals last year
– or 200 per hour: Activists condemn Wildlife Services, a division of the USDA, which says deaths necessary to protect farmers and public health; TheGuardian.com; March 2022

List and kill count of wild animals

According to *The Guardian* article, the list of the most-often killed wild animals killed in 2021 by the USDA's Wildlife Services division included:
- 1,028,642 European starlings
- 143,903 feral swine
- 75,351 northern pikeminnows (fish)
- 66,554 pigeons

- 26,969 Canadian geese
- 24,683 beavers
- 21,706 chestnut munias (birds)
- 63,965 coyotes
- 19,170 mourning doves
- 16,792 black vultures
- 15,198 brown tree snakes
- 15,096 red-winged blackbirds
- 12,670 house sparrows
- 10,775 black-tailed prairie dogs
- 10,589 scaly-breasted manias
- 10,162 ravens
- 9,961 zebra doves
- 9,627 brown-headed cowbirds
- 9,583 double-crested cormorants
- 9,003 white-tailed deer

Other wild animals killed

Other lists of animals killed by Wildlife Services include nutria rodents, mice, rats, racoons, wolves, bears, mountain lions, bobcats, foxes, groundhogs. Species killed "unintentionally" or intentionally include dogs, birds, squirrels, rabbits, turtles, and other animals.

M-44 cyanide cannisters

A situation often mentioned relating to the killing machine that is the USDA department of Wildlife Services is the 2017 case of 14-year-old Canyon Mansfield of Pocatello, Idaho. He had been covered by toxic powder and his dog was killed. They were out walking and triggered an M-44 cyanide cannister set up to kill animals like coyotes and foxes considered a threat to livestock.

Fruit and vegetable farms are safer

While it is clear that many animals killed to protect farmed animals, and more are killed to protect the crops grown to feed farmed animals, vastly far fewer animals are killed to protect fruits and vegetables. No carnivore animals need to be killed to grow fruits and vegetables.

Feeding vegans is safer

Some will argue that feeding vegans kills a lot of animals. When it is taken into consideration the numbers of wild animals killed in the processes involved with growing and harvesting feed crops for farmed animals, and the numbers of wild animals directly killed to protect the feed crops, and the wild animals killed to protect the farmed animals, it is clear that a vegan diet results in both far fewer animals killed, and fewer varieties of animals killed. Especially for polycrop farms.

Polycropping and organics

Less damage is done to wildlife by growing multiple crops on a piece of land (polycropping), rather than clearing a piece of land and only growing one crop (monocropping).

Organic farming – as in: not using toxic chemicals – is less harmful to wildlife and the environment than what is now being called "traditional" farming (which can involve a variety of pesticides, miticides, fungicides, synthetic fertilizers, and other chemicals.)

Support your local organic produce farmers.

Grow food and compost food scraps into garden soil

It is also less harmful and more sustainable to grow your own food in an organic garden that also uses your food scraps as compost – rather than to rely on stores and restaurants for every calorie.

My garden

With my garden, I haven't killed any animals. I have fencing surrounding the garden, and also have built portable cages I place over areas of the garden. I don't use insecticides or other chemicals in the garden. I used to grow more food than can be eaten in my household, and often gave away what I grew to friends and neighbors. The most crops I lose to animal damage are the bugs that attack maybe a tenth of the strawberries. No other fruit or vegetable I grow had any significant damage from insects or birds. There was a squirrel that took the small tomatoes. Lately, there haven't been tomatoes. Bees, wasps, butterflies, ladybugs, praying mantises, and other beneficial insects and other bugs used to be in my garden. Their global populations have plunged. Please read Haddad's book *Insects: A Silent Extinction.*

Central to dietary choices

"As ever more people make the ethics of harming animals central to their dietary choices, the number of wild animals killed by farming is an essential piece of information – a figure that could inform ethical calculations, guide consumer decisions, and shape environmental research. A precise figure, however, is surprisingly difficult to find.

In the United States, an estimated 40 million cattle, 120 million pigs, and 9 billion chickens are killed each year. (The numbers are hazier for fish, who tend to be counted by weight rather than individually, but the final tally is likely several times higher than that of chickens.)

The mortal toll and environmental burden of feeding those animals has pushed many people towards plant-based diets. Yet wild animals are killed in producing those plants, and their accounting is decidedly unrigorous."

– Brandon Keim, *The surprisingly complicated math of how many wild animals are killed in agriculture*; *Anthropocene Magazine*; July 2018

The little things play into the bigger things

Because of the environmental benefits of each species, killing wild animals for centuries to protect farmed animals and feed crops has been ruinous. Including in the distance as wildlife helps form and nurture headwater forests, wetlands, ponds, and other parts of the naturescape that play into the health of rivers, lakes, marshes, and the oceans, and the wildlife dependent on them – which then plays into the production of oxygen, and the formation of clouds, storms, and weather patterns.

In Nature, the little things all play into the bigger things – and into the air, water, food, and wildlife you need to survive.

The slaughter of wildlife continues

According to *Wildlife Damage Review*, in 1994 the government spent over $56 million of federal, state, and cooperative funds to kill 783,585 wild animals. Since then, the budget has greatly increased.

According to ProjectCoyote.org, in 2021, the government spent over $124 million to kill 1.76 million wild animals, including birds and mammals. It continues these killings, "in the midst of biodiversity collapse, and the climate crisis… Wildlife populations have plummeted more than two-thirds in the last 50 years. One million of the estimated eight million plant and animal species are at risk of extinction." The killings continue "despite peer-reviewed research showing that reckless slaughter of native carnivores causes broad ecological destruction."

Ranchers lease land from the government at low rates that do not make up for what the government spends to kill animals, build roads, and supply water. It equates to corporate welfare for ranchers.

Killing birds

In addition to killing wild animals for pastureland, the U.S. Department of Agriculture also is involved in killing millions of birds. They regularly use poisoned rice laced with DRC-1339 to kill grain-eating birds, such as grackles, red-winged blackbirds, ravens, magpies, vultures, and yellow-headed blackbirds. The killing doesn't stop there.

This is because predator birds – such as Cooper's hawks and prairie falcons that eat the poisoned birds – also die.

In the summer of 2010, people in the Midwest states were surprised to find tens of thousands of dead birds around their towns, until it was revealed the birds were found to have been poisoned by the U.S.D.A. to stop them from eating cattle feed.

Killing predator animals

Killing predator animals damages the cycles of native animal and plant life. When the predator animals are killed off, the animals they feed on are able to reproduce unchecked. This has resulted in large populations of animals such as mice, rats, gophers, squirrels, badgers, prairie dogs, skunks, possum, rabbits, chipmunks, deer, and raccoons. This can play into lack of nutrient spread, and an imbalance of plant life, such as by deer and other animals denuding the natural landscape in ways that then alter creeks, rivers, and lakes, and the life in them.

The U.S. Government's Animal Damage Control (ADC) programs attempt to control the populations of the smaller animals by sending out trappers to eliminate them. They kill the smaller animals by burning them (including by setting fire to their dens), by bludgeoning them, and by poisoning their young. The poisoned animals are then sometimes eaten by other animals, sickening or killing them.

As mentioned earlier, many smaller native animals are killed to prevent them from eating the crops grown to feed livestock, and to prevent them from creating nesting and dwelling holes or dens in the ground, which may result in livestock injuring their legs. A cow with a broken leg in a distant "grazing" field are too difficult to transport, and would be left there to die, and possibly shot to quickly kill them.

"Hunting season" – killing what predators would have

To help control the population of some animals, including deer, hunters are allowed to enter into controlled areas to kill a certain number of animals every year. With the bear, cougar, wolf and populations way down, and jaguars having been eliminated by the meat and feed grain industries, what those predators would have killed and kept in check, are left to reproduce.

When bows are used to hunt deer, about half of the deer escape with the arrow stuck in them. The deer then suffer slow, agonizing deaths.

"Wildlife refuges" are hunting landscapes: guns, arrows, and traps

"Trapping and/or hunting are allowed on more than half of the 540 U.S. National Wildlife Refuges.

According to the U.S. Fish and Wildlife Service, of 27 million people who visited refuges, 22 million came for wildlife observation, while only 1.2 million visited to hunt or trap animals."
– A Voice for Animals, 2005; VoiceForAnimals.org

The techniques used to kill off smaller animals, and to eliminate brush and other native plants used by larger animals, also harm the bird populations. Not only do birds die when they get caught in traps, and when they eat poison meant for other animals, raptors are poisoned – and many die – after eating poisoned animals.

Where did the birds go?

Several years ago while spending most of a day outside, I noticed an absence of birds. All day I watched for birds, but saw none. I was in an area with trees and open meadows. Recognizing the absence of birds, I did a dive into what was going on. What I found out was both a surprise and not a surprise. But disturbing, sad, tragic, and likely another issue of Nature most people are not away of. If they knew how much their every breath and calorie depends on the presence of birds, bees, and other wildlife, maybe they would be concerned. Maybe they would do something about it. Including planting trees, growing pollinator gardens, spreading wildflower seeds, and being part of the solution.

The global population plunge of birds is impacting all animal and plant life, and nutrient spread, and the formation of soil, oxygen, clouds, rain, streams, rivers, ponds, marshes, and lakes, and the pollination and presence of plants humans depend on for food.

Bird habitat continues to be destroyed, their foods poisoned, migratory flights disrupted by lights and buildings, and water polluted. The ability of birds to survive decreases every year. Insectivore birds are particularly in trouble. Humans can't survive without them.

Avian loss in the billions since 1970: one-in-four birds gone

"For the first time, researchers have estimated the volume of total avian loss in the Western Hemisphere – and it's not just threatened species that are declining. Many backyard favorites are also losing ground.

While birds remain everywhere, people are actually seeing far fewer of them than just 50 years ago, according to a new study (*Decline of the North American avifauna*; Science.org, Sept. 2019). It estimates that North America is home to nearly three billion fewer birds today compared to 1970 – that's more than 1 in 4 birds that have disappeared from the landscape in a mere half a century.

About 90 percent of the missing birds came from 12 distinct and widespread bird families, including warblers, sparrows, blackbirds, and finches. Common birds found in many different habitats – even introduced, ubiquitous species like European starlings – experienced some of the steepest drops. Feeder birds like the dark-eyed junco declined by nearly 170 million individuals, while white throated sparrows dropped by more than 90 million."

– Jillian Mock, *North America Has Lost More Than 1 in 4 Birds in Last 50 years*, New Study Says; Audubon Magazine, Audubon.org; Sept. 2019

Half of world's bird species in decline, many face extinction

"Half of the world's 10,000-odd bird species are in decline. One in eight faces the threat of extinction. This problem has been worsening for decades, which means scientists have been able to estimate roughly how many fewer birds are around today than, say, half a century ago. The numbers are startling.

There are 73 million fewer birds in Great Britain alone than there were in 1970. Europe has been losing about 20 million every year, says Vasilas Dakos, an ecologist at the University of Montpellier in France – a loss of about 800 million birds since 1980. And in the U.S., just shy of 3 million individual birds have disappeared in only 50 years.

'We are seeing a meltdown of bird populations,' says Ariel Brunner, director of BirdLife Europe and Central Asia, a conservation NGO. Loss of habitats, the rising use of pesticides on farms, and, yes, climate change – these are among the factors to blame. Even if you are not a birdwatcher, the loss of birds impacts you. Birds regulate ecosystems by preying on insects, pollinating plants, and spreading seeds – by excreting them after eating fruit, for example."

– Chris Maraniuk, *Bird Populations Are in Meltdown: Humans rely on birds to eat insects, spread seeds, and pollinate plants – but these feathered friends can't survive without their habitats*; Wired.com; June 2023

Plummeting birds across almost all habitats: tipping point species

"*The 2022 U.S. State of the Birds* report shows plummeting bird populations across almost all habitats while highlighting the need for further conservation efforts.

The Rufous Hummingbird, Great Sage-Grouse, Pinyon Jay, and 67 other birds in the United States are teetering on the edge of disaster, having lost at least half of their populations in the past 50 years. A report released by North American Bird Conservation Initiative (NABCI) calls these birds 'Tipping Point' species, on track to lose another 50 percent of their populations in the same time frame if conservation efforts do not improve.

While some duck, goose, and swan populations are exploding, more than half of all U.S. bird species are dwindling. Without further efforts to restore ecosystems under stress, the report paints a grim future for birds in a nation where climate-intensified natural disasters and human-caused habitat loss and degradation continue to worsen."
– Margo Rosenbaum, *More Than Half of U.S. Birds Are in Decline, Warns New Report*; Audubon.org; Oct. 2022

We are watching the process of the sixth mass extinction

"Protecting wildlife and biodiversity is something everyone should be concerned with. Once we save birds, we're going to save a lot of other species we share the Earth with.

We're basically watching the process of the sixth mass extinction."
– Peter Marra, director of Earth Commons at Georgetown University Institute for Environment and Sustainability; 2022

"Staggering decline" of bird populations

"Because birds are conspicuous and easy to identify and count, reliable records of their occurrence have been gathered over many decades in many parts of the world. Drawing on such data for North America, Rosenberg et al, report widespread population declines of birds over the past half-century, resulting in the cumulative loss of billions of breeding individuals across a wide range of species and habitats. They show that declines are not restricted to rare and threatened species – those once considered common and wide-spread are also diminished. These results have major implications for ecosystem integrity, the conservation of wildlife more broadly, and policies associated with the protection of birds and native ecosystems on which they depend.

Species extinctions have defined the global biodiversity crisis, but extinction begins with a loss in abundance of individuals that can result in compositional and functional changes of ecosystems. Using multiple and independent monitoring networks, we report population losses across much of the North American avifauna over 48 years, including once-common species and from most biomes.

Integration of range-wide population trajectories and size estimates indicates a net loss approaching 3 billion birds, or 29% of 1970 abundance. A continent-wide weather radar network also reveals a similarly steep decline in biomass passage of migrating birds over a recent 10-year period. This loss of bird abundance signals an urgent need to address threats to avert future avifaunal collapse and associated loss of ecosystem integrity, function, and services.

Slowing the loss of biodiversity is one of the defining environmental challenges of the 21st century. Habitat loss, climate change, unregulated harvest, and other forms of human-caused mortality have contributed to a thousandfold increase in global extinctions in the Anthropocene compared to the presumed prehuman background rate, with profound effects on ecosystem functioning and services. The overwhelming focus on species extinctions, however has underestimated the extent and consequences of biotic change, by ignoring the loss of abundance within still-common species and in aggregate across large species assemblages. Declines in abundance can degrade ecosystem integrity, reducing vital ecological, evolutionary, economic, and social services that organisms provide to their environment.

A total of 419 native migratory species experienced a net loss of 2.5 billion individuals, whereas 100 native resident species showed a small net increase (26 million).

Species overwintering in temperate regions experienced the largest net reduction in abundance (1.4 billion), but proportional loss was greatest among species overwintering in coastal regions (42%), southwestern aridlands (42%), and South America (40%).

Shorebirds, most of which migrate long distances to winter along coasts throughout the hemisphere, are experiencing consistent, steep population loss (37%). "
– *Decline of the North American avifauna*; Science.org; Sept. 2019

Nobody should be killing wild birds

A decrease in the bird population allows more insects to populate the land. To kill off the insects, the government and farmers use more chemical poisons. The pesticides then do even more damage to the bird population, create insects resistant to pesticides, destroy beneficial insects (bees, wasps, butterflies, praying mantises, and ladybugs and other beetles, etc.), and pollute the land and water, in a cycle that would naturally take care of itself if humans would not interfere.

Animal Damage Control Act of 1931: killing wolves and grizzly bears

"The ADC program, created by the Animal Damage Control Act of 1931, is greatly responsible for the virtual extinction of the grizzly and wolf in the lower 48 states as well as for putting the black-footed ferret, jaguar, black-tailed prairie dog, bald eagle, and other wild animals in, or close to, the endangered category. ADC reported it poisoned 1.8 million animals in 1991 and distributed thousands of pounds of

restricted-use pesticides to private individuals who poisoned untold numbers more. The U.S. Agency for International Development works with ADC to export ADC pest control practices and chemicals, including those banned in the U.S., to developing countries."
   – *Wildlife Damage Review*, 1997

Ranchers & Bureau of Land Management workers kill plants

Ranchers and BLM workers eliminate shrubs used for food and shelter by native animals, and then plant grasses to feed grazing cattle. This practice also is done other countries. It is globally damaging native shrub and grasslands, and reducing bird and insect populations.

Meat industry subsidies: welfare farming destroying biodiversity

By eliminating native plants that provide food and shelter for wild animals, by eliminating small animals that provide food for predator animals, and by killing all kinds of native animals, the government has had an enormous negative impact on the populations, life cycles, migrating patterns, and social structures of native animals of North, Central, and South America (many migratory birds travel from North to Central and South America). This is done at great cost to taxpayers to provide grazing land for livestock. It is welfare farming and it is destroying the biodiversity and ecosystems of the continents.

Sadly, because more and more cattle are being raised on other continents, the practices of killing wildlife to raise livestock are becoming the standard in other countries.

Functions of predator animals play into Nature's circle of life

Predator animals not only manage populations of large animals in the natural circle of life, they also improve conditions for small animals, fish, insects, and plants, and the soil, air, and water.

Wolves of Yellowstone

When wolves were reintroduced to Yellowstone National Park, willow trees, cottonwoods, and aspen trees began to grow more abundantly there. This was the result of elk and deer seeking the safety of higher ground away from the wolves. The new vegetation attracted other wildlife, including beaver. The banks of the rivers and creeks, which had been damaged by elk and deer, began to show more vegetation growth, which in turn improved the health of the rivers and creeks. This improved conditions for fish and amphibian populations. And the new trees have also attracted native and migrating birds to nest there and provide shade for certain varieties of plants to grow.

Animal agriculture: the "greatest threat to human welfare"

"Next to an all-out nuclear war, today's intensive animal agriculture represents the greatest threat to human welfare in the history of mankind."
– Farm Animal Reform Movement, FarmUSA.Org

Road building and wildlands and wildlife

Next to the cattle industry, road building has caused a tremendous amount of damage to wildlife and its habitat. When looking at a present-day map of any country, it is easy to see how roads have segmented wildlands. Wherever a road has been built, many acres of wildland have been paved over, killing and/or displacing many forms of wildlife, and often altering or destroying wetlands, creeks, streams, and rivers.

Serengeti National Park highway: damaging more wildlife

As people protested, the Tanzania government has been built a gravel truck highway through the Serengeti National Park, rangeland for the world's longest land animal migratory routes.

"Not only would we lose huge numbers of wildebeest who would simply die out, there would be the knock-on carbon effect. The Serengeti is the biggest carbon sink on the planet and if the wildebeest go, the grass burns, and we have a second environmental blow. The Serengeti is the world's best-studied ecosystem. Economically, ethically and environmentally, this mustn't happen."
– Andrew Dobson, Princeton University, speaking of Tanzania's plans to build a highway through Serengeti National Park

"The Great Migration" system of Nature in Africa

The Great Migration is a tremendous natural phenomenon involving gazelles, wildebeest, and zebra. While those are the main animals involved in the migration, many plants, animals, and insects are reliant on the migration. The predator animals, including cats, hyena, and crocodile feed off of the migrating animals. Vultures eat both the animals that die naturally, and the leftovers of animals killed by predators. Plants and soil organisms feed off of the urine, feces, and decomposing bodies of the animals. Birds and insects rely on the plants that grow from the soil nourished by the animals. Even aquatic life in the region's rivers and also birds rely on the insect life supported by the activities and life cycles of the animals. Interfering with the Great Migration could devastate the entire circle of life on, in, and around the African continent – and globally.

What's damaging African wildlife?

Hunting, poaching, human-spread diseases, and human encroachment has been devastating African wildlife for hundreds of years. It has sped up in the past several decades, including because of climate change, droughts, storms, the loss of food sources, and starvation. Some of the animals have been reduced by 90% – or more – in the past century. It would be extremely unwise to alter the landscape in such a way that would lead to the loss of more species on the African continent. A road with a continual flow of vehicles, including large trucks and service vehicles, and accompanying support structures and utilities causes more harm to wildlife.

"A dirt track may suffice today, the populations of these towns are only about half a million people each. Projecting forwards, say 50 years, and thanks to the new roads these towns will become cities of three to four million people each. The Tanzanians should not be assessing the impact of a narrow strip of road as we envision it today; it will not be a narrow strip in 30 or 40 years. A railway line will parallel it, and there will probably be a six-lane highway – in each direction. This, for certain, will kill the migration."
– Richard Leakey

Just as the livestock and feed crop industries should not be killing millions of wild animals every year, the road-building industry should be taking great measures to protect wildlife from harm.

Much of the roads that exist across the great planes, farmlands, and on land that was once rainforests of the world have been built to support the animal farming industry, and industries supporting the animal farming industry, such as the freed crop industry, and the meat, milk, and egg processing and shipping industries.

Humans and the animal kingdom: love, enslavement, and killing

"The love for all living creatures is the most noble attribute of man."
– Charles Darwin

"We have enslaved the rest of animal creation and have treated our distant cousins in fur and feathers so badly that beyond doubt, if they were to formulate a religion, they would depict the Devil in human form."
– William Ralph Inge

230

"The cruelties associated with the slaughter of the animal kingdom for human consumption, the pain, fear, and distress suffered by the animals in the entire process of being fattened for butchering, as well as the environmental disasters wreaked upon our planet through the meat industry, are very well documented, and should be understood by all who claim to be developing the enlightenment, or who wish to."
– Bodo Balsys

"As the crisis of feeding the world's population grows, breeding of animals for human consumption becomes less acceptable – out of compassion for the suffering of animals and the awareness that it is a grossly inefficient use of water and grain. A new relationship with the animal kingdom is part of our changing perception of the Earth. Animals are part of us, and part of our practice."
– Allan Hunt Badiner

"Every individual who eats flesh food, whether an animal is killed expressly for him or not, is supporting the trade of slaughtering and contributing to the violent deaths of harmless animals."
– Roshi Philip Kapleau

Grazing land and millions of acres of monocropped feed crops

According to National Geographic, about 26% of Earth's ice-free landscape is used for cattle grazing, and about 36% of land used to grow crops is used for feed crops for farmed animals. Only about 55% of the world's crop calories directly feeds humans. The rest goes to feed farmed animals.

According to NPR, in the U.S. alone, 644 million acres are used for livestock grazing, and 127.4 million acres are used to grow feed crops to fatten farmed animals.

Palm Oil

Decimating native habitat and wildlife, and causing climate change

In addition to feed crops being grown on and cattle ranching taking up far more than a billion acres of land around the globe, which decimates native habitat and wildlife, one plant cultivated to such extremes and in ways tremendously destructive to the land and environment are oil palms grown on massive plantations.

Over the past several decades, oil palm plantations are using up more and more land in tropical regions – where rainforest was killed.

Origins of oil palms, and the massive destruction of plantations

The oil palm is native to West African, and was brought to southeasts Asia during the colonial era. For thousands of years, palm oil has been used as a food, in cosmetics, and as fuel.

In the 1800s, palm oil became common as an industrial lubricant. By the late 1800s, palm oil was a chief export from west African countries.

As palm oil plantations started spreading into other countries, by the mid-1900s Columbia developed their palm oil industry into the leading producer in the Americas.

Palm oil plantations rapidly expanded in southeast Asia, with Malaysia and Indonesia becoming the world's leading producers, with Nigeria placing third.

From 1995 to 2015, oil palm plantations quadrupled in acreage.

In addition to deforestation, oil palm plantations have led to the destruction of damp peatlands that sequester carbon dioxide. When the peatlands are drained, they not only stop absorbing CO_2, but release enormous amounts of it into the atmosphere, contributing to global climate breakdown. This is especially so when millions of acres of peat are destroyed – which is exactly what has been happening.

The environmental destruction caused by the rapid spread of oil palm plantations in tropical regions has been happening as the countries importing the most palm oil are the U.S., Canada, European countries, and India.

Destroying the land, plants, and animals of the tropics for palm oil

"The rapid expansion of oil palm plantations has led to the destruction of huge areas of Indonesia and Malaysia's tropical rainforests, amplifying carbon dioxide emissions and destroying the remaining habitats of many endangered species.

The tropical forest species index has declined by 25 percent in the last three decades, according to the World Wildlife Fund. Globally, 300 million hectares of tropical forest were converted for non-forest use during the last 20 years of the twentieth century.

Between 2015 and 2020, the rate of deforestation was estimated at 10 million hectares per year. Several research studies have revealed that most of the world's palm oil plantations are within these converted hectares."

– Abhijit Mohanty, *World Environment: How our greed for palm oil is destroying tropical rainforests*; DownToEarth.org

Palm fruit oil and palm kernel oil, and their uses

Palm oil is a shade of red when it is raw, and it comes from the palm fruit. Palm kernel oil is clear and comes from the palm kernel.

Palm oil is found in everything from body lotion, shampoo, toothpastes, and cosmetics to breads, junk food, fast food, garbage snacks, instant noodles, some brands of peanut butter, chocolate, and ice cream. Palm oil and palm kernel oil have become the most widely-used oils in foods and cosmetic products. A high percentage of U.S. grocery store items contain palm oil.

Globally, more than 22 times more palm oil is produced than olive oil.

Simply because palm oil is in so many products doesn't mean it is safe, or healthy.

Clever ingredient wording used to hide palm oil

As many consumers have sought to avoid the use of palm oil, companies have gotten clever and started avoiding using the words "palm oil" and "palm kernel oil" on their products. They will use the terms vegetable fat, vegetable oil, oil, palmitate, glyceryl, palmate, octyl palmitate, elaeis guineesis, and hydrogenated palm glycerides.

Damaging to heart and cardiovascular system

Look into it, and you can find why it is good to avoid using products containing palm oil, including because heated palm oil is one of the most common fats that can damage your heart and cardiovascular system – triggering heart attacks and strokes.

Saturated fat of palm fruit and palm kernel oils

Palm oil is 50% saturated fat, and palm kernel oil is 80% saturated. In addition to heart disease, palm oil has been associated with arthritis, diabetes, inflammation, kidney disease, leaky gut syndrome, liver disease, and skin problems.

Rapid expansion of oil palm plantations causing great harm to Nature

"The rapid expansion of oil palm and other oil seed crops is causing significant environmental damage, leading to debates about their impact. Using advanced deep-learning techniques and satellite data, we created the first detailed global map of oil palm plantations in 2019, distinguishing between industrial and smallholder plantations with high accuracy. Our findings reveal the extent of these plantations in 49 countries, covering approximately 19.60 million hectares, with Southeast Asia being the main producing region."
– TheTreeMap.com/global-palm-oil

From the year 2000 to 2020, more than 25 million acres of forest were destroyed in Indonesia. At least a third of that was done to plant oil palms. Even when forest is lost because of the lumber industry, much of that then gets planted with oil palms or feed crops, or is used as cattle grazing land.

Millions of acres of tropical forests destroyed for oil palm plantations

"In Borneo, an island split among Brunei, Indonesia, and Malaysia, the palm oil industry caused roughly 40 percent of deforestation between 2000 and 2018, or roughly 6 million acres of forest loss. That's almost five times the size of Delaware.

When forests fall, so do hugely important ecosystems that influence the entire planet. The jungles of Indonesia and Malaysia are home to a stunning array of plants and animals, including orangutans, tigers, and the world's largest flower, the stinking corpse lily. Wet forests known as peatlands – many of which have been drained and replaced by

plantations – also store massive amounts of carbon, which can escape into the atmosphere when they're destroyed."
– Benji Jones, Vox.com, 2023

"Palm oil has been and continues to be a major driver of deforestation of some of the world's most biodiverse forests, destroying the habitat of already endangered species like the orangutan, Bornean pygmy elephant, Sumatran elephant, tigers, and Sumatran rhino."
– World Wildlife Fund, 2023

Borneo's forests: 90% destroyed

On Borneo the logging and farming industries have destroyed 90 percent of the forests. Those forests contained massive amounts of carbon, actively sequestered more, provided tremendous amounts of oxygen, held huge amounts of water, provided homes to a large numbers of wild animals, and helped rain clouds and soil to form, and the weather to regulate. Ridding the land of its native plants and animals has caused a huge release of both carbon and methane into the atmosphere, advancing climate breakdown and global warming, and helping to cause massive storms and flooding in other regions.

Top palm oil producing countries: massive environmental damage

The world's top palm oil producing countries are Indonesia, Malaysia, Thailand, Columbia, Nigeria, Guatemala, Papua New Guinea, and Honduras. In other words, oil palm plantations have been causing massive environmental damage around the globe.

"Oil palms now cover 11 percent of Indonesia's Sumatra, the world's sixth-largest island. In places like Sumatra, the industry's growth has also pushed some of the rainforest's wildlife species to the brink of extinction.

In Indonesia and Malaysia, many endangered wild animals are losing their homes to make room for oil palms, including elephants, tigers, proboscis monkeys, and Sumatran rhinos. These two islands are the only home for about 80,000 critically endangered orangutans and their habitats are under constant threat of deforestations due to oil palm plantations.

The loss of rainforests limits the ability to sequester a large amount of carbon. This ultimately destroys the rich biodiverse habitats. Large swaths of forest are burned down for the production of palm oil as this is the cheapest way. Clearcutting and burning of trees releases a large amount of carbon into the atmosphere."

– Abhijit Mohanty, *World Environment: How our greed for palm oil is destroying tropical rainforests*; DownToEarth.org

Oil palm plantations and logging: orangutans on road to extinction

Eliminating the rainforests to plant oil palms has placed the orangutans there on a road to extinction, expected to come to a dead end within decades.

We have no idea of the plants and animals we've sent into extinction

As so much of the forests have been destroyed, so are many types of life. Scientists don't go in and sample all of the forms of life of a forest before the forests are destroyed. We have absolutely no idea what we have lost. Plants and animals have been sent into extinction because of the spread of palm oil plantations around the globe.

Burning forests for palm oil plantations burns and starves wildlife

As I write this there are more forest fires burning on Borneo. Intentionally set to expand the oil palm plantations, the fires cause orangutans to flee, often to places humans live. Some people torture the orangutans for entertainment. Other people consider the orangutans to be pests and shoot them. Some of the orangutans are captured and sold into the exotic pet, circus, and zoo markets. Some of the orangutans die from burns and injuries suffered in the forest fires, and their babies die of starvation.

Don't use palm oil or palm kernel oil. Avoid foods and products containing them.

Massive monocropping operations damage wildlife

What is happening with the destruction of rainforests to expand the oil palm plantations is similar to what is going on with other wildlands as massive monocropping operations of a variety of crops – mostly feed crops – spread among every continent and on some islands.

Cattle and feed crop industries even more destructive than palm oil

From 2000 to 2022, the cattle and feed crop industries have deforested more than four times as much land as the oil palm plantations.

Want to do your part help stop rainforest destruction? Stop eating meat and dairy, and avoid foods and other products containing both them, as well as palm oil and palm kernel oil.

Humans are the only animals that extract oils and add them to food

According to a Nature.com article in 2020, the global demand for vegetable oils is projected to increase by 46% by 2050 (*The Environmental impacts of palm oil in context*). The thing is, humans are the only animal that extract oils from things, and adds the oils to their foods.

Adding oil to your foods is not a nutritional requirement. You easily get essential fatty acids by eating a variety of fruits, vegetables, sprouts, nuts, and seeds. There is oil in every cell of every plant. For instance, when you squeeze lemon into water, there is a sheen of lemon oil floating on the water.

Bottled oils are often heat-processed (even some with labels claiming it is "cold pressed"). The heating changes the chemistry of the oil. Oil is damaged more when the foods are cooked, grilled, barbequed, fried, or sauteed.

Food oils, including olive oil, contain saturated fats which contribute to inflammation and cardiovascular disease. For more information, see the site ForksOverKnives.com, which was put together by a variety of doctors, nutritionists, and nutritional bioscientists.

Palm oil: deforestation, biodiversity decline, greenhouse gasses

"Planted oil palm covers less than 5 to 5.5% of the total global oil crop area due to oil palm's relatively high yields. Recent oil palm expansion in forested regions of Borneo, Sumatra, and the Malay Peninsula, where 90% of global palm oil is produced, has led to substantial concern around oil palm's role in deforestation. Oil palm expansion's direct contribution to regional tropical deforestation varies widely, ranging from an estimated 3% in West Africa to 50% in Malaysian Borneo. Oil palm is also implicated in peatland drainage and burning in Southeast Asia. Documented negative environmental impacts from such expansion include biodiversity declines, greenhouse gas emissions, and air pollution."
– *The Environmental Impacts of Palm Oil in Context*, Nature.com; December 2020

The Domino Effect
of Humanity Living Unsustainably

If people would live more sustainably

If people would massively reduce the use of fossil fuels, follow a plant-based diet, grow and/or wild harvest some of their foods, support local organic produce farms, and adapt to more environmentally friendly lifestyles, it would make a big difference in what is going on in and around the oceans, and in and around the streams, rivers, and lakes of the continents and islands – all the way up into the mountains.

Fossil fuel use continues… destroying us

More and more people are exploring ideas of how we can sequester the CO_2 from the atmosphere to reduce global warming. This is good – especially if it involves planting trees and other plants, and protecting wildlife habitat. We also need to drastically cut down on the amount of fossil fuels we use, and to do so quickly.

In 2010, the U.S. was importing over $600,000 worth of petroleum every minute.

It hasn't improved.

The U.S. both imports and exports petroleum

Read this about how much petroleum was both imported and exported into and out of the U.S. in 2022:

"In 2022, the United States imported about 8.33 million barrels per day of petroleum from 80 countries. Petroleum includes crude oil, hydrocarbon gas liquids (HGLs), refined petroleum products such as gasoline and diesel fuel, and biofuels. Crude oil imports of about 6.85 million barrels per day accounted for about 74% of U.S. total gross petroleum imports.

In 2022, the United States exported about 9.52 million barrels per day of petroleum to 180 countries, and 4 U.S. territories. Crude oil exports of about 3.60 million barrels per day accounted for 38% of total U.S. gross petroleum exports. The resulting total net petroleum imports (imports minus exports) were about -1.19 million barrels per day, which means the U.S. was a net petroleum exporter of 1.19 million barrels per day in 2022.

The top five source countries of U.S. gross petroleum imports in 2022 were Canada, Mexico, Saudi Arabia, Iraq, and Columbia."

– U.S. Energy Information Administration, *Frequently Asked Questions: How much petroleum does the United States import and export?*; EIA.gov

In March 2022, people were surprised to learn the U.S. had been importing petroleum from Russia when the U.S. produces so much oil that it also exports. They learned this when Russia invaded Ukraine.

"The U.S. does indeed produce enough oil to meet its own needs. According to the U.S. Energy Information Administration, in 2020 America produced 18.4 million barrels of oil per day, and consumed 18.12 million. And yet that same report reveals the U.S. imported 7.86 barrels of oil per day last year.

That happens because of a combination of economics and chemistry. The economics are simple: overseas oil, even after shipping costs, is often cheaper than domestically-produced crude. That is because what oil people call 'lifting costs,' the cost of actually getting the oil out of the ground, are so much lower in some other countries. That, in turn, is down to a number of factors. Environmental and other regulations here play a part in that cost differential of course, but, contrary to what some would have you believe, they are far from the be-all and end-all affecting prices.

Land and lease prices are a big factor, as are labor and other costs."

– Martin Tillier, *America Produces Enough Oil to Meet Its Needs, So Why Do We Import Crude?*; NASDAQ.com, March 2022

The above article goes into other details about why the importing and exporting of petroleum is so off kilter.

About 40% of the large ships traveling the seas are carrying petroleum products, coal, and forms of natural gas. This fossil fuel business continues at a time when climate change, global warming, and ocean acidification increase, and wildlife populations are plunge.

Military-industrial complex: war lords, petroleum lords

It all plays into the military-industrial complex, and those who benefit from it. Count in the war lords and petroleum lords – who are often the same people.

The military of many countries are major polluters

With all of its many thousands of cars, trucks, planes, boats, and other fossil fuel-burning equipment, the U.S. military, and the military operations of other countries are major polluters.

Part of what many military operations do is work to secure the petroleum sources on the planet.

If they really want to work on protecting the people, which is also protecting Earth, they should be working to get off petroleum, off coal, off natural gas, and off nuclear energy, and work on restoring Nature. Imagine millions of military personnel around the planet doing that.

The U.S.: fossil fuels hog

In 2011, the U.S. was using about 25% of the petroleum being used in the world. By 2022, per person, the U.S. still ranked as the lead consumer of petroleum, using 20.3% of the world's consumption of petroleum.

Dietary choices play the major role in fossil fuels use

The leading way in which U.S. citizens use fossil fuels is through their dietary choices. In the U.S., that means: how much animal protein they consume, and what is used to produce each calorie of the protein.

U.S. citizens are eating more animal protein than other countries, and are the leading consumers of the massively energy-intensive product: beef, the flesh of slaughtered cattle.

U.S. raises millions of cows, and imports even more meat

In addition to the millions of cows being raised in the U.S., the country imports massive quantities of beef, seafood, and other meats to support the meat-heavy diets most U.S. citizens consume.

Food and fossil fuels

There is a direct correlation between the use of fossil fuels and the consumption of meat, dairy, and eggs.

As has been pointed out, and which most people don't seem to understand is, the number one use of fossil fuels relates to what humans eat.

Most of the food grown on every continent is fed to farmed animals raised for their milk and flesh.

Per person, the wealthier countries consume the most animal protein, and also use the most fossil fuels – including in relation to their food choices.

The solution

The solution to the highly-polluting, environmentally-degrading ecocide diet rich in animal protein is to switch to a plant-based diet, and one also based on locally grown fruits and vegetables. Especially organically grown foods. This will dramatically cut the use of fossil fuels, localize economies, and improve human self-reliance, nutrition, and health, and the air, land, and water, and situation of wildlife.

Sequestering carbon

While there is ongoing talk of sequestering carbon from the atmosphere by planting more trees, growing more plants, and using various plants as building materials, including by using bamboo, hemp, hay bales, and agricultural waste in construction, people seem to have forgotten Nature has a variety of ways of sequestering carbon, including through forests, seaweeds, and algae.

The creation of poop in humans and wildlife, including sea mammals and other sea life, also sequesters carbon. The average U.S. citizen poops about 320 pounds per year. That's about what an adult panda bear weighs.

Much of what humanity has been doing is destroying so many forms of wildlife that even the natural way of sequestering carbon through wildlife poop has been greatly reduced.

Wildlife populations decreasing, globally

All sorts of wildlife populations are decreasing. Reasons include loss of habitat, loss of food sources, increasing temperatures, pollution, environmental degradation, diseases, road building, and being killed by human activities, including urban sprawl, driving and other transportation, and, chiefly, because of human dietary choices.

Mountain gorillas in Africa's Virunga National Park, which are edging closer and closer to extinction, are experiencing respiratory infections spread to them by humans.

In Australia, the Tasmanian devils, which are another species decreasing in numbers, have been succumbing to a transmittable cancer that deforms their faces to the point they die.

In North America, a fungus has been spreading and killing millions of bats.

Each of these animals, gorillas, Tasmanian devils, and bats, play more of a role in the environment than people seem aware.

Wild animals support plant life

All wild animals help support the plant life in their natural environments. The plants feed off of the nutrients left by the droppings of the wildlife. The droppings are sequestering CO_2, and the plants are sequestering CO_2. The animals eat more plants, and then create poop, which helps build soil, and healthy plant life, and so the circle goes. The natural behavior of wildlife also spreads seeds and pollinates plants, which creates more plants, and more oxygen. The process plays into everything from the health of the soil, air, water, and the other life forms dependent on them – from the bottoms of the seas to the mountains.

Bats, mosquitos, and pollination

People may think of bats as scary little things that can get caught in a person's hair. How many people do you know who have had a bat caught in their hair – other than nobody, ever? Movies and books have also perpetuated a fear of bats, often portraying them as something frightening. In reality, next to bees and other insects, bats are important to people. Bats are one of the major pollinators – including of fruits people eat every day. Bats also consume an enormous number of mosquitoes, helping to keep their populations in check.

A fungus causing what has been named *white-nose syndrome*, was identified among bats living in a New York cave in 2008. The syndrome has quickly spread throughout parts of the United States east of the Mississippi, and into northeast Canada, killing entire colonies of bats. As the fungus is spreading and more bats are dying off in North America, there is concern about how some insect populations will increase, and spread of disease.

Bat colonies can also sequester CO_2, through their poop that often ends up in caves and other enclosed spaces.

A decrease in the population of bats is not a good thing, including because of how much CO_2 they sequester, and because bats are pollinators of and help to fertilize wild plants, which helps build soil, and sequester more greenhouse gasses.

Benefits of whale poo: sequestering carbon and fertilizing oceans

When researchers at Australia's Flinders University published a study detailing how much whale poop contributes to the sequestering of carbon, whale protection advocates pointed that out as one more reason why whales need to be protected.

Flinders University researchers recognized that the poop from sperm whales in the Southern Ocean contains significant amounts of iron. The nutrients in the whale poop are fed on by phytoplankton. When the phytoplankton die, the carbon trapped in their bodies sinks to the bottom of the sea.

Sperm whales in decline, and being killed

Sperm whales consume squid growing in the deep waters, and they poop in the surface waters where the phytoplankton live. One problem is there are only about 12,000 sperm whales left in the Southern Ocean, which is likely only 10% of the number of sperm whales that had been living there only two hundred years ago. The main reason for the drop in the population of sperm whales, and all other whales, is because of whale hunters. Increasingly, it is boat strikes, plastic pollution, other pollution, and nutrition issues.

"Why are the Russians killing these whales? They were using sperm whales for spermaceti oil, a high-heat resistant lubricating oil. One of the things that they were making with them was intercontinental ballistic missiles. So here we are destroying this incredibly beautiful, intelligent, magnificent creature for the purpose of making a weapon meant for the mass extermination of human beings. That's when it occurred to me that we as humans are insane.

From that moment on, the change in my life was that I never did anything again for people – I did it for whales and other creatures of the sea. So that pretty much puts us beyond the criticism from people – because when people disagree with what we're doing, I say: I don't care. Our clients are the whales, sharks, seals, fish, whatever. We don't give a damn what you think. Find me one whale that disagrees with what we do, and maybe we might reconsider. But until then we're going to do what we do."
– Paul Watson, Sea Shepherd Conservation Society, to *Earth Island Journal*; Autumn 2010

It is a travesty that whale hunting continues. However, global warming and ocean acidification pose even greater threats to whales. If the plankton keep dying off, there will be no whales. The krill populations will plunge, and so will those of krill-eating seabirds.

Of plankton and pteropods

What is happening with plankton is also happening with pteropods.

"Two groups of tiny, delicate marine organisms – sea butterflies and sea angels – were surprisingly resilient in the face of dramatic global climate change and Earth's most recent mass extinction even 66 million years ago.

Sea butterflies and sea angels are pteropods: abundant, floating snails that spend their entire lives in the open ocean. A remarkable example of adaptation to live in the open sea, these animals have thin shells and a snail food transformed into two wing-like structures that enable them to 'fly' through the water. (Pteropod comes from the Greek words for 'wing' and 'foot.')

Sea butterflies have been a focus of environmental change research because they make their shells out of aragonite. Aragonite is a form of calcium carbonate that is 50% more soluble than calcite, which other open ocean organisms use to construct their shells.

Because pteropod shells are susceptible to dissolving in more acidified ocean water, they have been called 'canaries in the coal mine.' – sentinel species that signal the impact of ocean acidification."

– Sentinels of ocean acidification survived Earth's last mass extinction; National Science Foundation, Oct. 2020

Ocean acidification is ruinous for shelled sea life

"Scientists have known for some time that shells of the tiniest sea life have been dissolving due to an increasingly polluted ocean.

Off America's West Coast, the snails are losing their shells at faster rates than previously thought.

'We did not expect to see pteropods being affected to this extent in our coastal region for several decades,' said William Peterson, an oceanographer at NOAA's Northwest Fisheries Science Center.

As fossil fuels are burned to create energy, the resulting carbon dioxide is released into the atmosphere, part of which settles into the world's oceans. As the ocean waters become more CO_2-enriched, the ocean's acidity increases, a process known as ocean acidification. At a certain threshold, the acidity can and does dissolve the snail's shells."

– Elizabeth Shell, As Pacific acidifies, 'sea butterflies' are quickly losing their shells; PBS.org, April 2014

"We do know that organisms like oyster larvae and pteropods are affected by water enriched with carbon dioxide. The impacts on other species, such as other shellfish and larval or juvenile fish that have economic significance are not yet fully understood."

– Richard Feely, co-author of study on pteropods, April 2024

Pteropods are winged underwater creatures that live in the polar and subpolar seas. These marble-sized creatures are the food for cod, herring, pollack, and salmon. Pteropods also have calcium carbonate shells that dissolve in acidic conditions.

Pteropods feed off tiny crustaceans, including plankton. Many types of fish feed off the pteropods. And larger creatures, including otters, penguins, puffins, eagles, gulls, albatross, terns, pelicans, tundra swan, snow geese, seals, sea lions, polar bears, grizzly bears, and black bears, feed on the fish.

If the pteropod populations plunge, so too will the populations of all of the marine life dependent on the pteropods, and so will the birds and land animals dependent on the fish they catch on shores, and from rivers.

Coral reefs, shell fish, starfish, jellyfish all being impacted

The demise of coral reefs and all sizes of shellfish, and echinoderms like starfish, urchins, sand dollars, are only parts of the problems waterlife face because of carbon dioxide, farming chemicals, and plastic, and urban and industrial pollution.

Ocean pollution

"We can see cities during the day and at night, and we can watch rivers dump sediment into the ocean, and see hurricanes form."
– Sally Ride, space shuttle astronaut

Fish, jellyfish, and other sea life suffer because of heavy metals pollution, such as mercury spewed by both coal-burning electrical plants and concrete kilns. The metals interfere with the normal function and growth of their body tissues. This includes how their bones grow, how their nerves function, and in their ability to reproduce.

Carbon dioxide from air pollution hovering over water is known to gather in the tissues of fish and interfere with their absorption of oxygen, asphyxiating them. They also die off because of a lack of oxygen in the water caused by liquid pollutants.

Fertilizers, nitrous dioxide: 300 times more damaging than CO_2

Fertilizers pose a double threat to marine life because fertilizers cause both water and air pollution. The fertilizers are made from natural gas extracted from Earth, which means they are nitrogen-based. When fertilizers are spread on farms, lawns, and landscaping, the fertilizers emit nitrous oxide, which is nearly 300 times more damaging to the environment than carbon dioxide. Synthetic chemical fertilizers run off from farmland during storms, and end up in rivers flowing into lakes and oceans. They also end up in ground water.

#1 use of toxic fertilizers: feed crops for farmed animals

The number-one use of synthetic fertilizer is for the growing of feed crops for farmed animals. No other crop comes close.

As mentioned, most of the food on every continent is grown to feed farmed animals, not humans.

Most of the water and fuel used to grow food is used to grow food for farmed animals.

Livestock consumes 70% of the grain, 80% of the corn, and 90% of the soy grown in the U.S. Other continents have similar rates.

In 2007 there were over 20 billion pounds of fertilizer and over 175 billion pounds of pesticides used on U.S. farmland to grow food for farmed animals.

33% of arable land on the planet is used for farmed animals

The farmland used to grow the food for farmed animals uses an astounding 33% of arable land on the planet.

As mentioned earlier, most or all of the wildlife that would have naturally lived on the land is displaced, or is killed off by farmers, ranchers, and government workers using guns, traps, poisons, flooding, fires, and/or torches.

Amazon rainforest and cattle

Seventy percent of the formerly forested areas of the Amazon rainforest are used for grazing cattle. These cattle compact and erode the land, damage creek and riverbeds, kill wildlife, and cause the extinction of species. The burning and clearcutting of all of rainforest land has released enormous amounts of carbon dioxide into the atmosphere, and has removed billions of trees that had been absorbing carbon dioxide while providing homes for varieties of wildlife. Removing the trees raises the temperature of the land, reduces the absorption of greenhouse gasses, and plummets the populations of wildlife.

The most meaningful action you can take to help the environment

"It's phenomenal to me that groups come out with lists like '20 Things You Can Do to Change the Environment' and will list 'drive a fuel-efficient car' and 'change your light bulbs,' but won't say 'eat less meat.' They are omitting the single-most powerful, most meaningful action you can take.

Let's say you take a shower every day and these showers average seven minutes. That's 49 minutes of showering a week. Let's say that's 50 minutes a week, with flow rates of two gallons a minute (which is very strong). At that rate, you'd be using 100 gallons a week for showering. That's 5,200 gallons a year to shower. It takes 5,214 gallons of water to produce one pound of California beef (according to University of California Agricultural Extension). You'd save as much water by not eating one pound of beef as you would by not showering every day for one year."
– John Robbins, author of *The Food Revolution*

Farmed animals produce 130 times more waste than humans

When it is taken into consideration how much pollution is created to grow the food for the farmed animals, that farmed animals in the U.S. alone produce 1.4 billion tons of manure each year by 9.8 billion farmed animals, or 130 times more waste than the country's human population; and the animals and their waste emit methane gas, which is over 20 times more damaging to the environment than carbon dioxide emitted from engines; and manure emits the greenhouse gas nitrous oxide, it is easy to understand that animal farming and the meat industry create more global warming gasses than all forms of transportation combined, including cars, trucks, motorcycles, ships, and airplanes.

Leading cause of global warming, pollution, acidification, species loss

In other words, it can't be more clear: hundreds of millions of farmed cows, sheep, goats, and pigs, and the billions of chickens, turkeys, and other animals raised for food, and all of the resources used to grow food for them, to slaughter them, and to transport, package, refrigerate, and cook the meat are easily the leading cause of global warming, deforestation, pollution, species reduction, and the acidification of the rivers, lakes, marshes, and oceans.

Excessive algae growth in oceans caused by animal ag

Fertilizers from farm and landscaping washing from the land, and nitrogen leaking into rivers from manure and urine pits cause excessive algae growth in the oceans. The floating algal blooms spread through miles of water and choke off waterlife, which results in "dead zones" where all forms of waterlife natural to that area cannot survive.

There are more than 400 dead zones – and more forming – in the world's seas. The dead zones take up vast amounts of water. They all are related to pollution, and more are now a mix of overly warm water mixed with industrial, residential, pharmaceutical, and other pollutants.

Dead zones

"With increasing temperature, and the fact that we are pouring huge quantities of nutrients into the oceans – things like agricultural fertilizers and sewage – we are seeing blooms of algae. Some of those algae are toxic. They affectively poison the environment. In other cases, the blooms are so thick that when those algae die, bacteria break down that material and use up all of the oxygen in the water, and of course that leads to suffocation of all of the other organisms living in the environment. And we are seeing the spreading of the so-called dead zones throughout the world's oceans and they seem to pretty much track where the heaviest concentrations of humans actually are."

– Dr. Alex Rogers, Scientific Director of IPSO and Professor of Conservation Biology at the Department of Zoology, University of Oxford

As mentioned earlier, in the Gulf of Mexico dead zones the algae and bacterial growth is directly related to the synthetic fertilizers and farmed animal waste washing into rivers flowing into the Gulf. There, the algal growth and subsequent dead zones help to increase water temperatures. The decrease in fish populations in and around these dead zones drives sharks closer to the coastlines, increasing the number of shark attacks on humans.

Millions of dead menhaden fish wash up in Texas

In June 2023, millions of dead menhaden fish washed up on the shores and floated in the surf off the coast of Brazoria County, Texas. The die-off was blamed on warmer water temperatures, with county park officials saying the warmer water holds less oxygen then cooler water. There is also a liquid nitrogen and propane plant on the coast of Quintana Island, and an enormous Dow Chemical plant that takes up hundreds of acres. It's a toxic waste zone.

Was it warmer water?

Was it chemical spills?

Was it a combination?

There in a county reliant on massive industrial chemical plants not far from Houston, who will ever know?

It is interesting that sea birds were not eating the fish. Perhaps the birds detected something.

Whatever happened off the coast of Brazoria County, millions of fish died. Bulldozers were brought in to shovel the billions of pounds of rotting fish into dump trucks.

Global blight experienced by oysters and clams

Dead zones are directly related to the global blight being experienced by oysters and clams throughout the world. The bacteria growing in the dead zones kill clam and oyster larvae. This process was first discovered at a facility in Newport, Oregon, in 2005. Just three years later, oyster and clam populations had plunged along the entire West Coast of North America. The number of larvae being killed by bacteria has been in the billions. East Coast waters are experiencing similar situations, as are other seas. Have you heard of it? You probably have not.

As people stare at TV, eat junk, and replicate commercial imagery

The corporate media is focused on whatever will keep people tuned in so the people watch commercials. What the corporate-owned media has found to successfully keep people tuned in are crappy TV shows, reality shows, day and night soap operas, game shows, and celebrity, political, and sports gossip, and talk shows of all those other people promoting their shows. Meanwhile, as people are staring at that nonsense while eating corporate foods, the environment is degrading.

Chesapeake Bay pollution caused by animal agriculture

"An analysis by the Chesapeake Bay Program found that agriculture – both livestock and crops – is the single biggest source of pollution in the bay, contributing 42 percent of the nitrogen, 46 percent of the phosphorus and 76 percent of the sediment in the troubled waterway."
– Monica Eng, Chicago Tribune reporter, The costs of cheap meat critics of factory farms say we pay a high price for low-cost food; September 24, 2010

"There are three major contributors to the poor health of our streams, rivers, and the Chesapeake Bay – nitrogen, phosphorus, and sediment.

High levels of nitrogen and phosphorus fuel unnaturally high levels of algae growth in the water, blocking sunlight from reaching underwater grasses that serve as food and habitat. When the algae die they are decomposed by bacteria that consume the oxygen in the water.

Too much sediment – tiny particles of dirt, sand, and clay floating in the water – turns the water cloudy, also blocking sunlight from reaching aquatic grasses. Oysters and other bottom-dwelling species can be smothered when that sediment finally settles to the bottom.

Excessive amounts of nitrogen and phosphorous come from fertilizers, wastewater, septic tank discharges, air pollution, and runoff from farms, cities, and suburbs. Excessive amounts of sediment are carried into our waterways from erosion and from construction sites.

45% of the nitrogen pollution in Chesapeake Bay is of animal waste and fertilizers that wash off agricultural land, or contaminate groundwater, polluting rivers and streams, and the Bay. 19% is air pollution. 17% is urban and suburban stormwater runoff. 16% is wastewater treatment and factories. 4% is from septic tanks.

The mainstem of the Bay and many of its rivers and streams are on the impaired waters list because of the reduced levels of oxygen and pollution that kill off fish, crabs, oysters, and aquatic life."
– What is killing the Bay?; Chesapeake Bay Foundation, CBF.org, 2023

Damage to the largest estuary in the U.S., Chesapeake Bay, has been vast and all forms of marine life existing there have been impacted. The tragedy is directly related to farming chemicals, farmed cow, pig, chicken, and turkey waste, industrial pollution, and to fertilizers used on lawns, sporting fields, campuses, and other landscaped areas. The water has become murky with algae, which blocks out light and allows for bacterial growth. Oyster, striped bass, and menhaden fish populations in the Chesapeake Bay have plummeted.

Main spawning area for striped bass rich with flesh-eating bacteria

Chesapeake Bay is the world's main spawning waters for striped bass, which are now often found to be starving and with bacterial infections eating away at their flesh.

Chesapeake Bay losing its fish, shelled creatures, and marine birds

The first to go were the oysters, which nearly vanished in the 1980s, and have never recovered.

The population of Chesapeake Bay blue crabs has plunged by 70% in less than twenty years.

The oysters and menhaden are natural water filterers, but even an abundance of them couldn't keep up with the amount of microscopic life growing in the bay and feeding on the farm pollution.

Dead zones spreading in world's seas

Dead zones of algae and bacteria overgrowth caused by farming and industrial pollution in the Atlantic, Mediterranean, Indian Ocean, and other saltwater bodies are impacting marine life in ways similar to what is happening in the U.S. Pacific, Gulf of Mexico, and Chesapeake Bay.

Animal waste and feed crop pollution causing algae overgrowth

In addition to the oceans, algae overgrowth caused by animal waste and feed crop farming chemical pollution is becoming a problem in the world's largest bodies of fresh water, and to swamps, saltwater marshes, and reefs. Again, ecocide because of one thing: animal farming.

Of melting ice caps, changing water alkalinity, and warming water

Waterlife throughout the world is in danger because of the melting ice caps, which results in fresher water at the poles, and saltier water near the equator. Marine life can exist only in water within certain ranges of salinity and temperature. If the temperature and/or salinity of the water are changed for too long, those species die off.

The more links in the circle of life that are lost, the weaker the others become.

Melting ice caps and plunging wildlife populations

"Melting sea ice will also take a toll on Arctic seals and walruses, which feed on tiny crabs and clams under the ice. Ringed seals, like polar bears, also rely on vast ice floes – in their case, to hide their young from predators. As the ice melts earlier each year, seal pups can be exposed to attack before they're able to defend themselves.

Newborn caribou, too, are imperiled by climate change. As Alaska's rivers thaw earlier, they create faster and more dangerous rapids that the young caribou must cross as they migrate from their nursing grounds."
– Natural Resources Defense Council, Campaign Update, Jan.–Feb. 2010

The melting ice caps are among several factors leading to the warming of the oceans.

With polar cap ice decreased, ocean water and the land heat up

Ice reflects the heat of the sun. Where the ice has melted, the ocean water and the newly bare land absorb the solar radiation heat that otherwise would have naturally been reflected back into space. This heat absorption delays and prevents the formation of new ice. The extent of ice melt of the polar caps is greater now than has ever been recorded.

"What is going on in the Arctic isn't something we can consider completely remote from ourselves. Actually, it's a fantastic barometer of what is going to happen in the rest of the world. The Arctic is ground zero for climate change and we're already pushing many species towards extinction. The key to preventing their loss is reducing our

greenhouse gas emissions – specifically carbon dioxide – to a level of 350 parts per million, or below. That is a level many leading scientists have called for to restore Arctic sea ice."
– Rebecca Taylor, Care for the Wild International, 2010

Even the lives of indigenous and other people living at great distances from the ice-covered poles depend on ice in the Arctic and Antarctica, and on glaciers, and permafrost terrain. These icescapes and frozen landscapes are collectively called the cryosphere. They are part of what would be helpful to understand – especially so people then take local and personal actions – and pressure governments – to help reverse the trend of global warming. That is, to live more in tune with Nature.

Warmer seas, sea rise, bigger storms, disrupted life, more methane

When warming temperatures cause ice to melt out-of-season, cause rain to fall instead of snow in the cold season, and cause permafrost and glaciers to melt, it disrupts life not only of animals, but also of bacteria and fungi, and of plant root systems, and the growth and survival of the water vegetation, including sea grasses, seaweeds, and algae. It's repetitive, as Nature repeats. Plants of both the land and water contribute to the production of oxygen, to the absorption of greenhouse gases, and to the food of wildlife that then play a role in healthy forests, grasslands, deserts, coastlines, rivers, lakes, marshes, oceans, and the cryosphere.

Pestilence, fungus, bacteria, more disease, and heat

When the permafrost melts, more mosquitoes and flies are present, increasing the risk of contagious diseases.

As the permafrost is depleted, bacteria and fungi grow faster and more abundantly, and they increase carbon dioxide in the atmosphere.

Nobody is more aware of the melting permafrost than the people who live near or in the Arctic. Where they once depended on the permafrost to keep their food cold, they are purchasing refrigeration units. And those are often run on electric generators using fossil fuels, which then puts more pollution into the air.

Northwest Passage accessible, seal pups exposed to predators

In September 2007, Arctic Ocean ice melted to the extent the Northwest Passage was accessible by boat for the first time in known history. In 2010, the rains continued into November, which is when it should have been snowing. Snow is needed for seals to survive as they keep their pups in dens, protecting them from the cold and predators, like bears, and foxes. In recent years, it rains later into the year.

In 2011, there was less ice on the North Pole than had ever been recorded. The trend has continued, with each year averaging worse.

The Bering Strait is now ice-free into December.

Polar bears are also losing the ice habitat needed to build their dens. Penguin are losing their terrain, are increasingly stressed, and having to search more for food, burning more calories, and their young suffer.

Permafrost melting causes the release of massive amounts of methane

The melt extends into the permafrost of the soil, releasing massive quantities of methane.

As mentioned, when the permafrost melts, the number of insects increases. It also brings birds to nest out of season, then lose their young to cold and/or starvation.

Greed and the exploitation of the newly-exposed land

Adding to this tragedy is that companies consider it as an opportunity to exploit the resources of this newly uncovered land. Petroleum and mining companies want access to this land to extract whatever they can find to sell into the world market.

As the ice caps melt, the interest in exploiting mineral, natural gas, and petroleum extraction there increases. The fishing industry is moving in, and shipping routes are changing, taking advantage of passageways formerly covered in ice. As polluting industries move in, destruction of the environment increases. Trash is tossed from ships and boats, air pollution increases from the engines spewing diesel exhaust, and the risk of oil spills becomes more likely along sensitive coasts, where wildlife struggle. Ships and military craft increase noise in the waters, disturbing hibernating bears, and hindering whale communication. As the acidity of the water increases, sound waves in the water become more prominent, and travel greater distances, harming sea mammals.

"Unchecked global warming and oil drilling in the Arctic put polar bears in double jeopardy."
– Andrew Wetzler, NRDC's Wildlife Conservation Project

Fossil fuels industries eager to destroy the land, water, and wildlife

Petroleum, natural gas, and mining companies are putting up a fight to be allowed to extract from the newly exposed landscape, and to build servicing facilities for their tankers and other equipment. Not only will allowing these industries in cause more damage to the environment, which impacts wildlife, it also raises the issue of emergencies – such as oil spills, fires, and other catastrophes in regions where, for months on end, sunlight is limited and temperatures dip well below freezing. Dealing with such emergencies is a problem even under the best of light and weather conditions where access is not deterred by intense winter storms or landscapes far away from supply sources. But, of course, industry leaders say they have emergency plans.

The petroleum disasters that took place in the Gulf of Mexico and with the Exxon Valdez stand as examples of what horrible things can happen in the waters, on the land, and along the coasts of the Arctic.

Later and shorter winters, more rain, thinner ice

Winter temperatures have been happening later and later in the Arctic, with the rainy season extending more and more, and the ice becoming thinner. Scientists have concluded the sea ice is half of what it was just three decades ago, and winter temperatures are continuing to rise.

None of these changes are good for the wildlife of the Arctic, or for anything dependent on healthy populations of wildlife there – or in distant regions through the interconnections of Nature.

Full spectrum of Arctic wildlife massively impacted

The warmer temperatures, declining snow packs, and thinner sea ice are not good for even the smallest forms of life in the Arctic, including the bacteria, fungi, algae, and microscopic plant life in the ice and frozen soil. Many of the smaller animals, such as krill, amphipods, copepods, pteropods, and other zooplankton, depend on the microscopic life for their food – especially the algae.

Without a healthful zooplankton population, the creatures that feed on the zooplankton lose their food source, and so forth and so on, up the Arctic circle of life food chain to the larger animals, including whales, seals, walruses, bears, foxes, and birds.

As I mention earlier, seals in the Arctic rely on the snow packs to form dens for their pups. Without enough snow, pups are left to the open, and become food for bears, raven, and foxes. While it is natural for some of the pups to be lost to the predators, a large number of the seal pups are being killed simply because the parents cannot build snow dens to keep them safe. Even when they can build pupping dens, the earlier spring temperatures are melting ice earlier in the year, which exposes pups to predators they wouldn't normally be exposed to. These changes are happening quickly. Within a few short years there may not be enough snow for any seals to build pupping dens.

Starving polar bears not giving birth, and turning to cannibalism

With fewer seals, more polar bears go hungry. Polar bears are already faced with a whole number of issues causing their populations to decline at alarming rates. As rains continue later into the winter, more snow melts, and ice packs turn to seawater; the polar bears are not gaining easy access to the variety of food sources they would normally consume.

One way polar bears obtain food is to stay near the holes in the ice maintained by seals and smaller whales – capturing some of the whales and seals as they come up for oxygen. Because they are traveling further to get food, are getting less of it, and more of them are spending more time on land rather than ice, polar bears are becoming thinner, which threatens their survival.

Recent behavior of polar bears displays a species in distress and desperation. Polar bears have been observed raiding geese nests and eating the eggs, foraging on grass, berries, and brush, scraping the lichen from rocks, and killing deer. Polar bears have been found swimming far from sea ice. More and more bears are found to be starving, with some turning to cannibalism.

If a female polar bear doesn't carry enough fat in her body, she has a problem getting pregnant, and going full term with a pregnancy. Those that do get pregnant but don't have enough fat reabsorb the embryo.

The dietary changes in polar bears are reflective of what is going on with other species living near the poles, including humans.

People in the cold zones: Less wild food, and having to buy food

People of the cold northern climates traditionally have consumed meat of wild animals, including from fish, whale, seal, moose, and caribou, and some birds, and the eggs of birds and fish. They also have harvested wild berries. They didn't depend on stores for meat or milk. Milk was not in their diet as raising cows in the Arctic isn't an option. In 2015 a gallon of milk was somewhere around $10 a gallon. As industrial activities, such as tar sand mining, and changes in weather have disrupted and reduced wildlife populations, the people of the North have found themselves in a predicament of what foods to eat, and especially how much money to spend on purchasing meat, eggs and other foods shipped or flown in from southern climates. How to store their food has also become an issue. While they normally would have depended on keeping their food in storage cut into the permafrost, as the permafrost has been melting, they depend more and more on refrigeration that functions, at least in the daytime, on generators run on fuel, which also costs money they weren't used to spending.

Warmer water: southern fish going north

As the sea ice melts, and Arctic waters become warmer, fish typically found in more southern waters are making their way north. Where the fish go, the fishing boats follow.

Fishing fleets also going north, and depleting sea of fish

People of the north are finding competition from fishing fleets, which are overfishing local stocks. Again, a problem caused by people who live far, far away who are burning fossil fuels, cutting down forests, grazing millions of cattle on every continent, raising billions of farmed animals that emit methane that warms the planet, etc.

CO_2 from distant human activity spreads and destroys globally

Sea ice naturally blocks the ocean from absorbing CO_2. As there is less sea ice, the oceans are absorbing more CO_2 created by distant ovens, kilns, and generators, and by airplanes, trains, trucks, cars, motorcycles, ships, and boats. Colder water also absorbs more CO_2, meaning the colder oceans are magnets for CO_2, but the more acidic water doesn't stay where it is, and makes its way to other oceans.

Even the slightest rise in ocean acidity damages shelled creatures

As the poles of Earth lose more sea ice, more CO_2 ends up being absorbed into the oceans. This raises the acidity of the oceans, and damages varieties of the shelled creatures dependent on more alkaline conditions.

Even the slightest rise in ocean acidity can be devastating for zooplankton, including pteropods and krill, and for coral reefs, crabs, lobsters, clams, barnacles, oysters, starfish, and for all of the forms of life dependent on those shelled creatures, and the life dependent on them, and so forth – out to the wider circle of life dependent on healthy oceans, including you.

In 2009, Canadian scientists found Arctic ocean water already acidic enough to damage shelled creatures. That is, not only acidic enough to interfere with the formation of their shells, but acidic enough to damage existing shells.

Waters off the coast of Oregon were some of the first to be studied for the acidity and found to be acidic enough to damage shelled creatures. More and more waters around the world are in similar stages.

The past several decades are just the beginning of what is a rapidly expanding problem with ocean acidity.

Shocking new lows for sea ice: polar bears in great danger

"The scientific evidence is now overwhelming that the polar bear and its Arctic home are in far greater danger than previously thought.

Over the past three decades, over a million square miles of sea ice – an area six times the size of California – have disappeared! Polar bears need that sea ice to survive. They range across hundreds of miles of it to find mates and hunt for the food they need for themselves and their young.

In recent years, the Arctic sea ice hit a shocking new low: there was less ice left than half the global warming models had predicted for 2050!

Most scientists are warning that nearly all of the polar bears' summer sea ice could vanish by 2040 – and some think it could happen much sooner.

No sea ice, no polar bears. It's that simple.

Polar bears are suffering the tragic effects of global warming. As their sea ice melts, they are being forced to swim further and further to ever-distant flows in search of food. Instances of polar bears drowning, starving, and resorting to cannibalism have all been documented."
  – Natural Resources Defense Council, The NRDC is helping fight the American Petroleum Institute, the National Mining Association, and other lobbies working to remove protections for the polar bear; 2011.

Arctic National Wildlife Refuge

There has been an ongoing fight taking place in Alaska to protect the Arctic National Wildlife Refuge from oil drilling to help provide some of the nearly six billion barrels of crude oil Americans consume every year.

"The accepted spill volume of the Exxon Valdez was 11 million gallons (262,000 barrels). But eyewitnesses say this was the low-end estimate. Five years after the spill, the State of Alaska released its independent analysis putting the spill at 30 to 35 million gallons."
  – Riki Ott, Hide and Leak: BP's Cleanup Is More Like a Cover up. Holding the Company Accountable Will Require Digging for the Truth; Earth Island Journal; EarthIslandJournal.org. Autumn 2010

"By far the most promising site for oil in America is the Arctic National Wildlife Refuge in Alaska."
  – George W. Bush, 16th Annual Energy Efficiency Forum; June, 2005

The "Energy Efficiency Forum" may as well have been called the "Greed Efficiency Forum." Or, the "Destroy Earth Forum."

Bristol Bay Alaska and gold, copper, and hard rock mining

A no less important fight is going on to protect the Bristol Bay Alaska headwaters region from gold, copper, and hard rock mining that would result in polluting the relatively pristine waters with toxic chemicals. This would kill millions of fish that are the food for bear and eagle populations, which in turn fertilize the surrounding forests through nutrient spread via there poo and eventual decomposing bodies.

Fewer of wildlife species isn't good for all patterns, and levels of life, including pollination, seed spread, soil organisms, root systems, soil water retention, soil building, and oxygen production, greenhouse gas sequestration, river and cloud formation, weather patterns, etc.

"For more than a decade, the threat of a huge, open-pit copper and gold mine has loomed over the heart of pristine salmon spawning territory in Bristol Bay, Alaska.

In January 2023, following a U.S. Army Corps of Engineers decision to deny a key permit for the mind, the U.S. Environmental Protection Agency releases a final determination that protects Bristol Bay waters from becoming a dumping ground for mine waste.

It is a hard-win victor by all those who have been finding for decades to stop the Pebble Mine project. Collectively, these decisions effectively spell victory in the decades-long fight to protect Bristol Bay.

Pebble Mine would directedly impact the world's greatest sockeye salmon run.

As proposed, the Pebble Mine would have entailed a mining pit over a mile long, a mile wide, and 200 meters deep, destroying nearly 3,500 acres of wetlands, lakes, and ponds, and 81 miles of salmon streams. And that only includes waters directly displaced by mine facilities, not the thousands more acres that would be fragmented, dewatered, and covered with dust from the mine.

The EPA cited its authority under Section 404(c) of the Clean Water Act in banning mine disposal in the salmon-filled waters. It took years of litigation to enable the Biden administration's EPA to follow the extensive scientific record and put forward these enduring protections"
– EarthJustice.org; 2023

"The State of Alaska and the Canadian mining company, Teck Cominco, want to create North America's largest open pit gold mine and a 896-square-mile mining district in the headwaters of Bristol Bay. At the same time, the Bureau of Land Management is trying to open 3.6 million acres of vital fish and wildlife habitat in the Bristol Bay Watershed to hardrock mining.

What most people don't know is that the hard-rock mining industry is the single largest source of toxic releases and one of the most destructive industries in America.

The proposed Pebble Mine may pose the greatest single threat to this area's salmon-bearing rivers. Similar open pit mines have devastated entire watersheds and surrounding fisheries throughout the United States and around the world. The Pebble complex would sit just 15 miles north of Alaska's largest body of fresh water, Lake Iliamna. If the mine is allowed to open, this lake could easily be contaminated with cyanide as well as with heavy metals, including mercury, arsenic, and selenium. Waters from this lake eventually drain into Bristol Bay. This whole region is considered to be more sensitive than the Alaska National Wildlife Refuge."
– Bristol Bay Alliance, 2007

"The proposed Pebble Mine may be the worst corporate assault on American's natural heritage that no one's ever heard of. Global mining giants – including Anglo American and Rio Tinto corporations – would gouge one of the world's largest open-pit mines out of Alaska's incomparable Bristol Bay wilderness.

My friends at the Natural Resources Defense Council call it the worst project they've ever seen – and they've seen hundreds of them. That's because this colossal mine would be built at the very heart of the headwaters of our planet's greatest wild salmon river systems: the Kvichack and the Nushagak.

Nothing like this place exists anywhere else on Earth. It is a remnant of American wilderness as it used to be, the kind of mythic landscape that Norman MacLean had in mind when he famously wrote: 'Eventually, all things merge into one, and a river runs through it.'

Make no mistake: an open mining operation in Bristol Bay could destroy some of Alaska's most sensitive habitats, and permanently contaminate a true wildlife Eden.

The mine would be absolutely huge in scale: with a gaping pit wide enough to line up nine of the world's longest cruise ships and deep enough to swallow the Empire State Building.

The industrial infrastructure would include transmission lines, 86 miles of shipping roads, and the dredging of Cook Inlet – home of the endangered Cook Inlet beluga whale – for a new deepwater shipping port.

And it is right here – in the heart of this American Eden – that foreign mining giants want to excavate their 2,000-foot-deep Pebble Mine. This monstrosity will spew some 10 billion tons of mining waste, laced with toxic chemicals (including cyanide, sulfuric acid, arsenic, and other toxic chemicals), that must be held back forever by massive earthen dams up to 50 stories tall – all in an active earthquake zone.

The biggest of five dams needed to hold back billions of tons of toxic waste would be one of the largest in the world – dwarfing even China's Three Gorges Dam!

The mine would sit in an earthquake-prone area near the Lake Clark fault, a 135-mile tectonic zone, and just 125 miles north of the site of the infamous 1964 earthquake: the most powerful earthquake in North American history.

In the event of a quake, billions of tons of contaminated waste could be release into the streams and habitats of Bristol Bay. But it wouldn't even take such a massive release to irreparably damage this unspoiled ecosystem.

That's because the plan for Pebble Mine also includes the permanent destruction of over 60 miles of salmon habitat, barriers to migrating fish, and the extraction of 70 million gallons of fresh water a day. And that may only be the beginning.

Salmon are acutely sensitive to pollution. An increase of as little as two parts per million of copper dust in the water can affect their ability to naturally navigate and spawn.

Sacrificing this living heritage in order to enrich a few foreign corporations would be an absolute travesty – one that no American should tolerate.

The Pebble Mine is an environmental disaster waiting to happen. If it pollutes the Kvichak and the Nushagak River systems, it will take down not only the world's greatest sockeye salmon fishery, but also the awe-inspiring ecosystem that depends on it.

Bristol Bay is home to orcas and beluga whales, wild moose and caribou, and one of only two populations of freshwater harbor seals in the world. Here you can find river otters, wolverines, porcupines, red fox and mink – all of which are intimately connected, directly or indirectly, to the health of Bristol Bay and its salmon.

The companies behind the Pebble Mine include some of the world's most powerful corporations: Britain's Anglo American and Rio Tinto.

According to a report commissioned by Nunamta Aulukestai and the Renewable Resources Coalition, Anglo American's history is littered with one toxic disaster after another: water contamination in Zimbabwe, mercury pollution in Nevada, high lead levels in children living near a mine in South Africa. The list goes on.

As for Rio Tinto, it has left a trail of pollution and destruction that spans the globe: from Bolivia to Indonesia, from Papua New Guinea to the United States. The *New York Times* reported that the Grasberg Mine – a Rio Tinto joint venture in Indonesia – left nearly 90 square miles of wetlands 'virtually buried in mine waste.' According to the Environment Minister documents, 'almost all fish have disappeared.'"

– Robert Redford and Frances Beinecke, Natural Resources Defense Council; savebiogems.org/bristolbay; June 2011

Alaskan coast eroding

"The Alaskan coast is expected to erode at a rate of 45 feet each year due to rising sea levels and stronger storms, leaving coastal communities and wildlife vulnerable to intense storm surges and flooding. Warmer waters pose a risk to Bristol Bay's salmon, the linchpin of this complex ecosystem, which need cool waters to spawn and incubate their eggs. A drop in Bristol Bay's legendary salmon runs will in turn jeopardize grizzly bears, a rare population of freshwater harbor seals, and other wildlife in the region that depend on salmon for food."

– Natural Resources Defense Council, Campaign Update, Jan.–Feb. 2010

Warmer atmospheric temperature = intense storms and floods

The changing ocean and atmospheric temperatures and the acidity of the water are playing roles in the accumulation of more water vapor in the global atmosphere. This is accelerating the occurrence of torrential storms, such as those that caused record flooding in the American Midwest in June 2008, and again in 2011, and in Korea in 2011, continuing on to 2023. Massive floods have taken place everyplace from Spain to Mexico to Vermont to India, to Australia, to Nigeria, to the Middle East, and so forth and so on. It is the new normal.

In October 2023, New York City experienced torrential downpours that caused flooding unlike anything ever seen there. They should expect this to become more common, and to worsen.

Excessive water in atmosphere is also because of deforestation

Other causes of the excessive accumulation of water in the atmosphere include the destruction of millions of acres of mountain forests and rainforests, which would normally hold water; clearing millions of acres of land of natural habitat on every continent to plant massive forests of genetically engineered timber, and plant huge monocrop plots of feed crops for billions of farmed animals; and the covering of billions of acres of land with driveways, roads, highways, bridges, parking lots, and parking garages to support continuing growth of motor vehicle culture, which paves over paradise.

Rising seas

Many low-lying countries, including Bangladesh, Guyana and Suriname, and Belize, and island nations like the Maldives, the Marshall Islands, Tuvalu, and Kiribati, are susceptible to rising seas within the next several decades. Likely sooner, rather than later this century.

Where will hundreds of millions of people move to?

There is the issue of certain nations having to abandon their land as it is taken over by the seas. Would the nation still be a nation, even if they moved to another land? What country would allow them in as an established nation? Would they still own the rights to resources extracted from their former land, even if it sits beneath water?

Many coastal regions are experiencing seas overtaking at least parts of their land, if even for a season. An increasing number of coastal villages and graveyards have been washed away. Water tables and land used to grow crops have become saline.

New Orleans and Hurricane Katrina

As it was broadcast on TV in 2005, New Orleans and surrounding areas experienced storm surges that overtook towns, washing away large numbers of homes and businesses, and flooded many areas, including most of New Orleans. It is clear the petroleum and natural gas industries have done great harm to natural barriers – including islands and marshes – that would have softened storm surges, and reduced damage. While what happened to the levees remains a controversy, there is no doubt the damage to the coast done by industry over the decades played a large role in the severity of the damage done by the storm swells relating to hurricane Katrina.

America's Gulf Coast is sinking as oceans also rise

There are an increasing number of areas along the Gulf Coast sinking beneath seawater, or where sea water is rising from the ground and forming puddles that get increasingly larger.

By continually depositing sediment, the Mississippi River took thousands of years to build up the land where New Orleans and the surrounding areas sit. The Mississippi has been so harmed by various projects meant to tame the river for industrial uses and for flood control it no longer flows in its natural patterns.

More and more beach closings: polluted water

Those of us who live in coastal areas are becoming aware of the damage being done to the oceans. In 2022 there were nearly 8,892 beach closings on U.S. state and territory shores triggered by pollution.

"Eighty percent of pollution to the marine environment comes from the land. One of the biggest sources is called nonpoint source pollution, which occurs as a result of runoff. Nonpoint source pollution includes many small sources, like septic tanks, cars, trucks, and boats, and larger sources, such as farms, ranches, and forest areas. Millions of motor vehicle engines drop small amounts of oil each day onto roads and parking lots. Much of this, too, makes its way into the sea.

Some water pollution starts as air pollution, which settles into waterways and oceans. Dirt can be a pollutant. Top soil or silt from fields or construction sites can run off into waterways, harming fish and wildlife habitats.

Nonpoint source pollution can make river and ocean water unsafe for humans and wildlife. In some areas, this pollution is so bad it causes beaches to be closed after rainstorms.

More than one-third of the shellfish-growing waters of the United States are adversely affected by coastal pollution."

– *What is the biggest source of pollution in the ocean?*; National Ocean Service of the National Oceanic and Atmospheric Administration; OceanService.NOAA.gov, 2023

In 2021, 33% of coastal beaches in the U.S. had at least one notification of polluted water. The pollution can be anything from industrial waste, farmed animal waste, leaking septic systems, stormwater runoff, pet and wildlife waste, military and industrial or recreational boat and cruise ship pollution, malfunctioning wastewater treatment plants, broken sewer lines, overflowing sewage systems, and harmful algal blooms that are often the result of lawn, landscape, and farming chemicals, and farm animal waste. Or, a combination of those pollution sources.

Beach closings are becoming common everywhere from the Middle East and Asia to Australia, from South and Central and North America to Europe and Mediterranean and Africa, and among islands throughout the world.

If you live near an ocean, there likely is an organization in your town or city fighting to keep the ocean and beaches clean. Many of these groups need volunteers, including for cleanup days. Join in.

Suntan lotion pollution: hazardous to fish & suffocating coral reefs

In Hawaii, so much suntan lotion ends up in the ocean water near tourist beaches that it becomes unhealthful for people to swim in, and hazardous for sea turtles, monk seals, coral reefs, and other marine life, including sea grasses, fish, and marine birds.

"Research shows the harmful chemical oxybenzone transferring from swimmers to fragile coral reefs.

According to concerned consumers at the Environmental Working Group, a nonprofit organization that rates sunscreen safety based on published scientific literature, oxybenzone is readily absorbed by the body, lingers for weeks on the skin, and in the blood, and may disrupt hormone production.

It's not just humans that are imperiled by oxybenzone, however; it's also the environment that suffers significantly as a result of contamination from sunscreen that contains this chemical. This was confirmed in a study published in the scientific journal *Chemosphere*.

Water and sand samples from Hanauma Bay in 2017 were done in order to measure the concentration of oxybenzone in the environment… Sunscreen pollution from a single day's contamination can linger in the bay for more than two days."
– Matt Alderton, *Sunscreen Pollution Threatens Hawaii's Hanauma Bay*; Treehugger.com; Nov. 2021

Scientists testing the sand and water from Hawaii's Hanauma Bay found pollution from suntan lotions reaching concentrations high enough to threaten wildlife and coral reefs. The study found, besides swimming, suntan lotions get into the ocean from beach showers.

Suntan lotions often contain the endocrine disruptor chemical oxybenzone, identified as having antiestrogenic and antiandrogenic properties, which can mimic, block, or interfere with hormones.

Starting in 2021, it became illegal in Hawaii to sell over-the-counter suntan lotions containing oxybenzone. The state also put a daily limit on visitors to Hanauma Bay. Other beaches, including Waikiki Beach, typically have a sheen of suntan lotion floating on the water.

Plastic pollution in the rivers, lakes, and oceans: a global problem

One of the main pollutants in rivers, lakes, and oceans around the worlds is plastic waste.

"Plastic is a material that the Earth cannot digest. Every bit of plastic that has ever been created still exists, except the small amount that has been incinerated, and has become toxic air and particulate pollution.

In the environment, plastic breaks down into smaller and smaller particles that attract toxic chemicals. These particles are ingested by wildlife on land and in the ocean, and contaminate our food chain.

Harmful chemicals leached by plastics are already present in the bloodstream and tissues of almost every one of us, including newborns.

Consumption of single use and disposable plastics has spiraled out of control. They are used for seconds, hours, or days, but their remains will last hundreds of years.

Unlike glass and metal, recycling plastic is costly and does not stem the production of virgin plastic product. Most of our plastic waste is landfilled, downcycled, or exported to other countries.

Patches of plastic pollution cover millions of square miles of ocean in the North Pacific and in the North Atlantic. Scientists expect to find similar accumulation areas in the remaining oceanic gyres. There is no known way to clean up the plastic pollution in the oceans: the plastic particles are very small and circulate throughout the entire water column. The amount of plastic pollution in the oceans is expanding at a catastrophic rate.

We believe that the term "marine debris" (used consistently by the plastics industry) is euphemistic, vague, and ineffective. We are here to talk about plastic pollution in the oceans, plastic pollution on land, and plastic pollution in our bodies. Let's call it what it is."

– PlasticPollutionCoalition.org

According to the American Chemistry Council, the U.S. produced 35.7 million tons of plastic in 2018.

In 2022, according to Plastic Oceans International, humans produced over 380 million tons of plastic. About half of that was for single-use plastic, including from packaging, food utensils, and sanitary and medical materials. Almost all plastic is derived from petroleum and natural gas. It is another example of the damage the fossil fuels industries have done and are doing to the environment and life on Earth.

As plastic became a popular material used in manufacturing an increasing number of products, it also grew into an enormous problem for the environment and wildlife – and human health.

Microplastics are in your body tissues and blood

As you are reading this, there are microplastics in your body. They got there through the foods you eat, the polluted air you breath, water you drank, and through your skin. They can help to trigger various health problems. Because there is so much plastic being used in modern society, you will gather more plastic in your body.

"Scientists discovered microplastics in human blood: About 77% of the people who were tested were found to have microplastics in their bloodstream.

Microplastics are small pieces of plastic – anything that's less than 5 millimeters in size – that break off from larger pieces of plastic. The fact that they're so small makes them dangerous: they're able to permeate tissues and get stuck in our organs, finding their way into places that bigger pieces of plastic can't access.

Microplastics may act as endocrine disruptors, meaning they behave like hormone in our bodies, confusing our regulating hormones into thinking there's too much or too little of that hormone, or blocking the effects of our own hormones. When our hormone balance is off, it could lead to fertility issues, blood sugar imbalances, metabolic issues, hypothyroidism, and hyperthyroidism, autoimmune diseases, growth problems, and much more.

Microplastics may contribute to cancer. 'This is purely conjecture, but knowing that microplastics are so small and can land on our cell membranes, they could cause issues with our cells' ability to regenerate healthy, new cells,' says Dr. Philip Kriakose of the Henry Ford Cancer Institute. 'Our bodies are very dependent on our ability to repair damage. Our cells are constantly regenerating and we're dying a thousand deaths every day, but we don't realize it because our genes are often able to suppress those abnormal, cancer-causing cells. But if microplastics are interfering with cell regeneration, it may cause a higher incidence of cancer, and other diseases."

– *Microplastics Have Been Found In The Human Bloodstream*; Henry Ford Health; HenryFord.com; April 2022

"The question isn't how are they getting into our bloodstream, the question is how could they not be? Microplastics have been found at the top of Mount Everest, and on the ocean floor. They've been found in a large, remote ice cap. They've been found in the placenta of fetuses. They're everywhere. We're living in a bubble of plastic – our homes, clothes, cars, beds, cosmetics, cups, food, drinks, bottles, and pacifiers – plastic is in everything."

– Philip Kriakose, M.D., hematologist and medical oncologist, Henry Ford Cancer Institute; 2022

Plastic pollution is found across the planet

From the inner cities, to remote forests, meadows, prairies, and deserts, and on top of the highest mountains, plastic pollution can be found across the planet. Throughout the coastal areas of the planet, on beaches in the middle and bottom of the oceans, and in areas far from major cities, tons of plastic pollution gathers on the sand, rocky shores, gets tangled among coral reefs and underwater rock formations, and mixes in the bottom sands.

There are bits of plastic of all sorts floating in the oceans, rivers, and lakes throughout the world. A 2006 study by the U.N. estimated each square mile of ocean had as many as 46,000 pieces of various sizes of plastic floating in it. It's only gotten worse.

Plastic bits are mistaken for food by seabirds, turtles, and other marine creatures, causing their death due to blocked digestive tracts. Other plastic materials, including fishing nets and plastic bags, end up strangling birds, turtles, otters, seals, and other animals that get entangled in the plastics. So much plastic has accumulated in the oceans there are large, swirling pools of plastic trash.

The Great Pacific Garbage Patch: a plastic pollution nightmare

Among the largest collection of plastic stews in the oceans exists in the Pacific halfway between California and Hawaii. It is known as the Great Pacific Garbage Patch. Some material pulled from the floating ecological disaster has been identified as being from plastics manufactured as early as the 1940s. Similarly old plastics have been found on the shores of North, Central, and South America, and New Zealand, India, and other areas.

Even when thrown "away" at a great distance from an ocean, plastic can impact far away marine life, and it can do so for many decades.

Because plastic lasts so long, the simple act of allowing plastic trash to enter lakes, rivers, marshes, or the ocean can result in the death of the fish, turtles, birds, whales, dolphin, and other animals living in the future. That is, if they do.

Plastics become more toxic as they accumulate environmental toxins

"I often struggle to find words that will communicate the vastness of the Pacific Ocean to people who have never been to sea. Day after day, Alguita was the only vehicle on a highway without landmarks, stretching from horizon to horizon. Yet as I gazed from the deck at the surface of what ought to have been a pristine ocean, I was confronted, as far as the eye could see, with the sight of plastic.

It seemed unbelievable, but I never found a clear spot. In the week it took to cross the subtropical high, no matter what time of day I looked, plastic debris was floating everywhere: bottles, bottle caps, wrappers, fragments. Months later, after I discussed what I had seen with the oceanographer Curtis Ebbesmeyer, perhaps the world's leading expert on flotsam, he began referring to the area as the "eastern garbage patch." But "patch" doesn't begin to convey the reality. Ebbesmeyer has estimated that the area, nearly covered with floating plastic debris, is roughly the size of Texas.

... Hideshige Takada, an environmental geochemist at Tokyo University, and his colleagues have discovered that floating plastic fragments accumulate hydrophobic, that is, non-water-soluble-toxic chemicals. Plastic polymers, it turns out, are sponges for DDT, PCBs, and other oily pollutants. The Japanese investigators found that plastic resin pellets concentrate such poisons to levels as high as a million times their concentrations in the water as free-floating substances."

– Charles Moore, *Trashed Across the Pacific Ocean, Plastics, Plastics, Everywhere; Natural History*; 2003

Disrupting the circle of life

A decrease in any variety of smaller types of waterlife is damaging to seabirds, bears, and other animals reliant on healthful populations of marine life.

When the bird and bear populations suffer, so to do the forests. The nutrients collected in the bodies of the birds, bears, foxes, and cougar and bobcat feeding from oceans, marshes, rivers, and lakes naturally end up as nutrients in the land, feeding the plants that then build soil, produce oxygen, sequester greenhouse gasses, and are homes to other wildlife.

Marine carbon and nitrogen isotopes: nutrients spread by animals

Marine carbon and nitrogen isotopes are two beneficial atomic nutrients brought into the forests by animals who eat fish. Even animals living hundreds of miles away from the oceans feed on fish that make their way upstream to spawn. The nutrients in the fish then end up as soil nutrients.

When the birds, foxes, wild cats, and bears don't get the food they need, their populations suffer. This has been happening for many decades. The impact of it is increasing, and is increasingly obvious.

Marine birds are dying, abandoning nests, lacking food

Large numbers of sea birds are dying, and many are abandoning their nests because they aren't finding the food they need to feed their young.

Killers of marine life: human activities

Pollution, global warming, water acidification, ocean noise from shipping and fossil fuel "exploration," injuries from boats and ships, and costal development are some of the reasons marine life is decreasing.

Fish farmers kill an increasing number of sea birds to protect the fish being kept in pens floating in the open oceans and lakes.

The birds are also being poisoned by industrial waste and by excessive amounts of bacteria caused by farming and industrial pollution putting marine life out of balance.

Marine birds dying after ingesting plastics

Plastic is a main reason for marine life death. An increasing number of bird carcasses are being found with large amounts of plastic in their digestive tracts. Including on islands far from human culture.

Bears: hungry, underweight, weak bones, abandoning cubs

What is going on with the marine birds is also happening with bears. Many bears are underweight, many are found to have weak bones, and others are simply abandoning their young. Bears that are underweight don't have babies because the embryo dies. As mentioned earlier, bears are also turning to cannibalism as their natural food sources are vanishing. The varieties of issues proving fatal to bears include industrial pollutants, plastics, overfishing, loss of habitat, desertification, and global warming.

Mountain headwaters forests being ruined: cattle, logging & sports

Mountain forests are where many of the rivers start that empty into the oceans. The forests are being damaged or eliminated by development, ski resorts, and the accompanying hotels and communities, "off road" sports, and also logging, road building, and the spread of cattle culture. The forests are also being damaged by pestilence and fires related to global warming. The loss of forestland is damaging the headwaters of rivers, the flood plains, and other wetlands, and the wildlife dependent on them.

Old growth being replaced with GMO trees & weak forests

Much forestland has been damaged by the logging of old-growth forests, and the planting of non-native trees, GMO trees, and timber trees planted too closely together. With all of the logging, there have been streams and rivers covered, and logging roads built, and this has altered the plant life. If the trees are planted too closely, it causes too much

269

shade, and native plants don't grow. Without native plants and the type supporting local native wildlife and migrating species, those animals don't show up, or can't survive. So, with that, the timber industry has damaged not only the streams and rivers, but, with the elimination of rivers, including by damning them, the ability of fish to return to their spawning waters is cut off. With the lack of fish that had otherwise been there for thousands of years, certain other wildlife also don't and can't exist there. Without those animals eating the fish and then transferring the nutrients into the forest through the animal defecation and, ultimately, the dying animals bringing nutrients into the forest, the forests suffer, the soil degrades, and with the non-native trees being so close together, the ground is not kept cool, and is not kept moist by undergrowth. The trees become networks of fuel for forest fires.

Fewer animals = depleted forests and weak trees

The depletion of wildlife populations is also impacting the forests because, as mentioned, the loss of wildlife is depleting the forests of nutrients, deep into the ground where fungus and bacteria help the root systems transfer nutrients into the plants and trees. When the forests don't get the nutrients they need, the immune systems of the trees weaken, making the trees susceptible to beetle infestations, rot, and fire. These are conditions accelerating specifically because of climate change, and drought. By the summer of 2008, California had already experienced more fires than for a typical entire year. That was at a time when salmon runs in the California rivers had also been lower than ever. It's largely only gotten worse.

Save the redwoods: Redwoods Rising

There has been some activity with groups restoring forest stream and river beds, digging out the fill used to create forest roads, and this has been bringing some fish back into the rivers. Wildlife is also returning to those small areas.

Save the Redwoods is a group doing excellent work. Sending them financial donations is a wise thing. They need all the support they can get. (Access: SaveTheRedwoods.org.)

"Redwoods Rising is an ambitious, large-scale project to restore areas damaged by historical logging in the globally significant forests of Redwood National and State Parks. These parks are home to almost half of the world's remaining protected old-growth redwood forests, which store more carbon per acres than any other forests on Earth.

They also safeguard imperiled salmon and trout, and rare creatures such as marbled murrelets and the endangered western lily.

However, despite the ecological riches and stunning beauty, these forests are far from pristine. Approximately two-thirds of the parks' 120,000 acres of redwoods bear the scars of industrial-scale commercial logging – some of which took place as recently as the 1990s. Logging not only took away huge, old trees, it also left behind heavily damaged streams and hundreds of miles of old, failing roads and stream crossings.

These forests will not recover on their own in the foreseeable future. We need help.

Redwoods Rising unites Save the Redwoods League, California State Parks, and the National Park Service to restore these previously logged forests."

– SaveTheRedwoods.org

Habitat fragmentation, fewer animals and plants, and more pollution

Animals outside of the heavily forested areas face some similar and some other intrusions, threats, and fatal interactions, including habitat fragmentation, pollution, invasive species, global warming, wildfires, drought, loss of food sources, decreased and degraded terrain, altered landscapes, roads and traffic, cattle ranching, people killing wildlife to protect farmed animals and feed crops, and also pipelines, and drilling and mining operations.

Biomagnification of toxins in wild animal food chain wreaking havoc

The predator animals, including bears and other mammals, living on and near the ice caps are also under threat from a silent danger biomagnifying in their food chain. Because the predator animals are at the top of the food chain, they collect pollutants in their bodies that had been in the tissues of the smaller creatures they ate. Although they live far from industrial society, the body tissues of these creatures have been found to contain fire retardants, pesticides, plastic softeners, chemical dyes, perfluorinated compounds used to make Teflon, other industrial chemicals, and pharmaceutical drugs. Marine birds, forest birds, seals, foxes, bears, whales, and fish living in the southern and northern regions of the planet have all been found to contain these chemicals. It is wreaking havoc on their hormone levels, on their immune systems, on their livers, on their bone and nerve structures, and on their birth rates.

271

Fire retardants in polar bears

One of the most common chemicals found in polar bears is a fire retardant used in furniture, blankets, mattresses, carpeting, and plastics, and in cell phones, televisions, and other electronics. These chemicals are known to disrupt thyroid function and hormone balance, impair mental abilities and motor skills, and to alter brain development. Bears are being found with weakened immune systems, and the milk of lactating bears has been found to contain enough of these chemicals to jeopardize the health of cubs. These problems are directly attributed to the pollutants the polar bears are accumulating in their tissues by consuming fish and other creatures living in or near polluted oceans. With fewer than 23,000 polar bears left in the wild, It is a quadruple tragedy they are facing with chemical pollutants, loss of food sources, plunging birth rates, and melting ice caps.

At one time the main threats humans posed to polar bears included only hunting and deforestation.

The existence of polar bears on the planet may soon come to an end.

As I'm writing this paragraph in 2023, everything is pointing in the direction of extinction for wild polar bears – probably within 20 years.

Oceans warming deeper and deeper, magnifying problems

"Notably, all ocean basins have been experiencing significant warming since 1998, with more heat being transferred deeper into the ocean since 1990. To date, the ocean contains 90 percent of the heat from human-induced global warming, and the year 2022 was the warmest ever measured for the global ocean."
– Global Climate Change; NASA.gov; 2023

Oceans takes up about 71% of the planet's surface area, and they are in deep trouble. The year I'm writing this paragraph, 2023, the oceans have been breaking records for heat. The ocean surface has especially warmed around Florida, and in the equator area of the Pacific, as well as the Indian Ocean, and between South America and Africa, but also in the north oceans.

In March and April 2023, the global ocean temperature jumped two-tenths of a degree. That has been the fastest rise on record.

Warming ocean temperatures means warmer clouds that hold more water, and create intense rainstorms that destabilize farming, flood towns and cities, damage roads, spread pollution, and kill wildlife.

Scientists from around the globe are speaking out about the state of the oceans, and trying to make governments and corporations take notice. They have been trying to get humanity to take action.

"Ninety percent of global warming is occurring in the ocean, causing the water's internal heat to increase since modern recordkeeping began in 1955.

Heat stored in the ocean causes its water to expand, which is responsible for one-third to one-half of global sea level rise. Most of the added energy is stored at the surface, at a dept of zero to 700 meters.

The last 10 years were the ocean's warmest decades since at least the 1800s. The year 2022 was the ocean's warmest recorded year, and saw the highest global sea level."
– Global Climate Change; Climate.NASA.gov; 2022

That was what NASA said in 2022. Then, 2023 happened, with the ocean surface heating faster than ever on record – it was of global warming conditions not expected to happen for decades.

"It's been about seven years since the last El Nino, and it was a whopper. The world has warmed in that seven years, especially the deeper ocean, which absorbs by far most of the heat energy from greenhouse gasses, said Sarah Purkey, an oceanographer at the Scripps Institution for Oceanography. The ocean heat content, which measures the energy stored by the deep ocean, each year sets new record highs regardless of what's happening on the surface.

Since the last El Nino, the global heat ocean content has increased .04 degrees Celsius (.07 degrees Fahrenheit), which may not sound like a lot, but 'it's actually a tremendous amount of energy," Purkey said. It's about 30 to 40 zettajoules of heat, which is the energy equivalent of hundreds of millions of atomic bombs the size that leveled of Hiroshima, she said."
– Seth Borenstein, *Earth in hot water? Worries over sudden ocean warming spike*; Associated Press; April, 2023

That was April, 2023. Then, the oceans got even warmer.

"Since April, the global average daily surface temperature of the Earth's oceans (excluding the polar regions) has remained at record levels, which is simply far too warm for the time of year. For example, according to Copernicus analyses, daily average maritime temperatures had already reached 20.94 degrees Celsius on July 19.

In addition, record surface water temperatures have persisted in the North Atlantic. In June, they averaged 0.19 degrees Celsius, the warmest of any Copernicus record for that period."
– Jeanette Cwienk, *Land and sea surface temperature hotter than ever before*; DW.com; August, 2023

And, the ocean heat continued to rise into the summer of 2023.

Earth had its warmest summer, ever. Again.

"Earth had its warmest August and June-August period on record; fifth consecutive month of record-high global ocean surface temperature.

Globally, August 2023 was the warmest August in the 174-year NOAA record. The year-to-date (January-August) global surface temperature ranked as the second warmest such period on record. According to NCEIs Global Annual Temperature Outlook and date through August, there is a 95% probability that 2023 will rank among the two warmest years on record.

Four continents – Asia, Africa, North America, and South America – had their warmest August on record. Europe and Oceana each had their second-warmest August on record. It was the warmest August on record for the Arctic. For the fifth consecutive month, global sea surface temperature hit a record high for the month, and overall August 2023 set a record for the highest monthly sea surface temperature anomaly of any month in NOAA's 174-year record.

Temperatures were above average throughout most of South America, Africa, Asia, North America, the Arctic, and Oceana. Parts of southern North America, central South America, western and central Africa, central, southern, and eastern Asia, and northwestern and eastern Oceania experienced record-warm temperatures this month. Sea surface temperatures were above average across much of the northern, western, and southwestern Pacific, as well as the Atlantic and the Indian Ocean. Record-warm temperatures covered nearly 13% of the world's surface this August, which was the highest August percentage since the start of records in 1951.

August 2023 set a record for the lowest global August sea ice extent on record. Globally, sea ice extent in August 2023 was about 550,000 square miles less than the previous record low from August 2019. Sea ice extent in Antarctica saw its fourth consecutive month with the lowest sea ice extent on record."

– Assessing the Global Climate in August 2023; National Centers for Environmental Information, NCEI.NOAA.Gov

Scientists are baffled by the jump in ocean heat

What has been going on this year with the ocean warming so quickly is freakish and baffling. Scientists can only guess what will happen because of it.

As the water becomes warmer toward the poles, it brings more people in contact with animals that formerly had little to no contact with humans. The result is hunters are killing more animals in places they had not been hunted. On top of this travesty, fishing fleets are traveling more northward and southward toward the poles, staying longer, and removing more fish and other waterlife, and leaving less of it for the wildlife that depend on it for survival. They are even taking one of the smaller sea creatures, krill, to supply the market for health food stores selling krill oil as an "omega 3 supplement" (a better choice would be for people to eat raw fruits and vegetables, hemp seeds, germinated seeds, and sprouts). The killing of marine life for human consumption remains a threat to sea creatures, and the wildlife that depend on those creatures.

"When a man wantonly destroys one of the works of man we call him a vandal. When he destroys one of the works of god we call him a sportsman."
– Joseph Wood Krutch

The worldwide fishing industry is ruinous to marine life

The worldwide fishing industry is playing a major role in destroying the oceans.

Fish species are becoming rare or extinct in regions where they were common just decades ago.

Every type of sea turtle is endangered.

Millions of sharks are killed every month.

Whales are still being killed.

People are taking sea corals to sell to the tourist industry, to the fish tank industry, and as decorative items.

Massive fishing operations continue to set billions of hooks every year to capture large fish. What they end up doing is killing sea life of all sorts, including sea mammals and birds.

It is estimated that 25% of the sea life captured by fishing fleets is not acceptable, and these dead or dying creatures are tossed back into the water. Some are kept to sell to the pet food, farm animal feed, and fertilizer markets.

"We're finally going to get the bill for the Industrial Age. If the projections are right, it's going to be a big one: the ecological collapse of the planet."
– Jeremy Rifkin, *World Press Review*, December 30, 1989

Predatory fish biomass only 10% of what it was in pre-industrial times

"We estimate that large predatory fish biomass today is only about 10% of pre-industrial levels.

We conclude that declines of large predators in coastal regions have extended throughout the global ocean, with potentially serious consequences for ecosystems."
– Ransom Myers and Boris Worm in *Nature*, 2003

Fishers are removing keystone species, and reducing gene stability

"The questions of how the impacts that we are having on the oceans will affect human lives and how climate change link to the oceans will affect human lives is a very complex one.

If we are talking in terms of fisheries, poor management of fisheries means that we don't reap the harvest in terms of food that we should be reaping from the oceans. We are actually extracting far too much fish. We are damaging the environment they live in. And also the fact that we are often removing keystone species – top predators that actually structure the environment – means we are causing really profound changes in the way many of these ecosystems are operating. The result of that overall is that one, the systems tend toward systems that produce organisms which are far less useful to us, like jellyfish, for example. The other issue is that those systems become far less able to cope with shocks. They become less stable. And that has really quite profound consequences for us and wider earth system terms, if you like.

Climate change affects are going to be extremely serious and it's interesting, when you think, many people talk about this in terms of what will happen in the future. Yes, my children will see the effects of this, or their children, or their children. While actually we are seeing very severe effects of climate change already. And we have been seeing that since the late 1970s – certainly for coral reefs. And this manifested in this form of this mass coral bleaching, which had never been seen before. At least, had never been recorded in human history. So, we are already there, really. And now, we are in the realms of just how serious are we going to allow this to actually get.

I always like to picture this almost as though alien invaders have arrived from outer space. They've killed off 19% of all of our coral reefs. They've exterminated more than 90% of many of the fish species that we depend on. And so on and so forth. What would be the response? Well, it would be a huge coming together across the world to take some sort of action against these invaders that are destroying our planet. Well, actually, we are doing this, already. So what this really demands is really global action to really take serious action in terms of reducing carbon dioxide emissions. And to stop to think about how we

actually draw carbon dioxide out of the atmosphere, because levels are already dangerously high – certainly for some marine ecosystems."
– Dr. Alex Rogers, Scientific Director of IPSO and Professor of Conservation Biology at the Department of Zoology, University of Oxford

Global ocean life collapse is on the way

A global study authored by 14 marine biologists that was published in the November 3, 2006 issue of the journal *Science* concluded that unless humanity makes enormous changes in the way they live and in what they eat, the populations of the world's fished species will collapse by about 2048.

The study considered evidence from all of the world's 64 large marine ecosystems. They found 91 percent of native species suffered from a 50 percent decrease, and 7 percent were extinct. Continued overfishing as well as coastal land development, habitat destruction, and air and water pollution are to blame. The study concluded that nearly 29 percent of fished species have collapsed (defined by being below 10 percent of historic highs). The study said fish populations are rapidly decreasing and losing entire functional groups.

The study authors factored the oceans will not be able to recover from the decline of so many species. The study authors wrote, "Our analyses suggest that business as usual would foreshadow serious threats to global food security, coastal water quality, and ecosystem stability, affecting current and future generations."

Many scientists throughout the world voiced their opinions in agreement with the study.

"We suggest that in the next decades fisheries management will have to emphasize the rebuilding of fish populations embedded within functional food webs, within large 'no-take' marine protected areas."
– Daniel Pauly, *Science;* 1998

Many scientists and environmentalists and organizations working to protect plants, animals, landscapes, forests, the air, headwaters, creeks, rivers, lakes, oceans, and everything about wild Nature have known for decades that things were going downhill. They have tried and tried to make a difference. They were ignored by industry and governments, in the name of greed. Decades of opportunities to reverse the environmental plunge and the loss of species have been lost. It has been happening as people watch TV, pay attention to celebrities, shop for stuff they don't need, put themselves in debt they should have not put themselves in, eat corporate foods, toss "away" more and more plastics and other trash, and contribute to the demolition of Nature.

Massive trawling nets dragged across the oceans, deeper and deeper

As mentioned, massive trawling nets are being dragged across the ocean floors at deeper and deeper levels to capture fish once abundant, but becoming sparse or nonexistent in places where they had existed not too many years ago. Many nets get caught on underwater rock or landscape formations, or on ocean trash and sunken boats, and are then abandoned as "ghost nets." This garbage continues to kill as fish, sea mammals, and marine birds get tangled in them. Through fishing, and the pollution they leave behind, massive fishing operations cause a destabilization of sea life biodiversity, extinguishing populations reliant on others to survive, and leaving other groups of ocean life destabilized.

The equivalent of killing every animal and plant in a forest

Deep-sea trawling is the equivalent of killing every bird, animal, and bug in a forest during a hunt for several hundred deer. Many of these massively destructive fishing expeditions operate on government subsidies that help with fuel and equipment, and are protected by laws formed to guard not the oceans or sea life, but the profits of the fishing industry.

Every human is subjected to dangerous chemicals

"For the first time in the history of the world, every human being is now subjected to contact with dangerous chemicals, from the moment of conception until death."
– Rachel Carson, *Silent Spring*, 1962

Pharmaceutical drug pollution

Additional damage is being caused to waterlife by the hundreds of millions of pharmaceutical drugs taken every day and that end up in the rivers, streams, lakes, marshes, and oceans.

According to the Centers for Disease Control and Prevention, 130 million Americans use prescription drugs every month. Not only are the drugs unsafe for the environment, many have proved to be unsafe for people.

Even when taken as prescribed, the vast majority of pharmaceutical drugs pose serious health risks. And when they do bad, they can do very bad, such as nerve damage, strokes, heart attacks, organ failure, and a variety of cancers and birth defects. They pose the same risks to the wildlife that ingest the drug-saturated water.

Estrogen is one pharmaceutical drug that ends up in the groundwater, aquifers, rivers, ponds, lakes, marshes, and oceans, resulting in more and more amphibians and fish having both female and male sex organs.

Pharmaceutical pollution may play a role in human health issues, including mood swings, stress, birth defects, blood sugar problems, learning disorders, early or delayed puberty, lower sperm counts, and pregnancy difficulties, including miscarriages and still births.

Even the way pharmaceutical drugs make it to market is rife with problems in the form of errors, tarnished manufacturing processes, contaminated packaging, dose errors, and quality control, and also cleverly altered studies, known risk cover-ups, government payoffs made, and lies told with the goal of making money for the pharmaceutical stockholders.

Many pharmaceutical drugs have to be recalled once they do make it to market. In some form, they end up in the environment.

Recalled drugs

In 2022 there were 912 types of drugs recalled from the market, and often because they were even more unsafe than even their warning labels detailed.

"Scientific advances have brought scores of new drugs in recent years. In the U.S., one major agency – the FDA – is responsible for making sure drugs they approve are safe and effective. Yet there were more than 14,000 drug recalls in the last 10 years, according to FDA statistics. That averages out to nearly four drugs recalled per day."
– Robert H. Schmerling, MD, *Drug recalls are common: Does that mean our medications are unsafe?*; Health.Harvard.edu; March 2023

"On average, about 4,500 drugs and devices are pulled from U.S. shelves each year. The recalled products have U.S. Food and Drug Administration approval and in many cases, are widely ingested, injected, or implanted before being recalled. Although the FDA may identify concerns regarding the safety of a drug, it is the responsibility of the manufacturer to initiate and execute a recall."
– DrugWatch.com

Where do these recalled synthetic chemical drugs end up? Eventually in the soil and water, or they are incinerated into the air, causing more pollution in the air, soil, and water – and the pollution ends up as residues in the soil, water, wildlife, and people.

Toxic foods, lack of exercise, health problems, and drugs

As the American way of consuming junk foods and an animal protein-rich diet while leading sedentary lifestyles spreads to other countries, so too does the occurrence of obesity, heart disease, diabetes, and other health conditions related to consuming junk foods and animal protein-dominant diets. This has led to a global increase in the number of people taking synthetic chemical prescription drugs to treat the health problems they get from unhealthful foods and lack of exercise. The drugs end up in the waters of the world because the drugs are urinated away, or expired and unwanted prescriptions are flushed down toilets.

Hospitals, nursing homes, long-term care facilities, testing labs, and the military often dump unused and expired drugs into sinks and toilets. As the chemical drugs dissolve into the waterways they wreak havoc on waterlife.

Pharmaceutical drug pollution altering fish, frogs, and other animals

"A few dramatic examples of wildlife harmed by drug contamination have been discovered previously, including male fish being feminized by the synthetic hormones used in birth-control pills, and vultures in India being virtually wiped out by an anti-inflammatory drug given to the cattle on whose carcasses they feed. Inter-sex frogs have also recently been found in urban ponds contaminated with wastewater.

Tom Bean at the University of York and colleagues, showed that the common antidepressant fluoxetine, at the low levels expected in the environment, led starlings to feed less often during the key foraging times of sunrise and sunset."
– Damian Carrington, *Drugs flushed into the environment could be cause of wildlife decline*; TheGuardian.com; Oct. 2014

There are many unknowns relating to how pharmaceutical pollution impacts wildlife. One reason is because the drugs aren't tested on wild animals, or on what the drugs do to plant life, and microorganisms in the soil and water. There is no way a drug company could afford to test their drug on every form of life. So, what happens to wildlife contaminated with pharmaceutical drugs is all a guessing game, mysterious, or an unknown.

One thing is certain: Pharmaceutical pollution alters wildlife.

"With rising living standards increasing the demand for livestock production and the growing and aging human population reliant on healthcare systems, pharmaceutical contamination of the environment is increasingly of concern. Although environmental contaminants such as pharmaceuticals are just one of the many stressors faced by free-ranging species (e.g., along with habitat destruction, climate change, and disease), with 41% of amphibian species, 21% of reptiles, 13% of birds, and 27% of mammals listed as threatened with extinction by the

International Union for Conservation of Nature, it is important that we understand the potential of pharmaceuticals to affect populations."
– Do Pharmaceuticals in the Environment Pose a Risk to Wildlife?; Society of Environmental Toxicology and Chemistry, SETAC.OnlineLibrary.wiley.com

Animal farming industry uses tremendous amounts of drugs

The quotation from the Society of Environmental Toxicology and Chemistry may as well have been more honest and said that the animal farming industry is using a tremendous amount of drugs on the farmed animals, and those drugs end up in the environment, and in wild animals, and in the people who eat the meat, dairy, and eggs. And the human population of the planet is eating an increasing amount of junk food. The diets of humans eating an increasing amount of animal protein and junk foods then get ill, got to doctors, and get drugs to treat their diet-induced maladies. Then, those drugs end up in the environment, and in wildlife. All of this is what is going on.

Little is known about drug hazards to wildlife

The Society of Environmental Toxicology and Chemistry goes on to say:

"Given the potential for pharmaceuticals to have therapeutic effects, side effects, or unexpected toxicity in nontarget wildlife, it is noteworthy just how many drugs are licensed for use and how little we know about hazard to wildlife. In the United States alone, the U.S. Food and Drug Administration (USFDA) has approved over 1,600 animal drug products and 20,000 prescription drug products for human use, which include one or more of the approximately 4,000 different APIs. A little over half (51%) of the drugs approved for use in veterinary medicine by the USFDA are also approved for use in humans. Notably there are no routine regulatory requirements for industry to perform tests in wildlife species for human or veterinary medicines."
– Do Pharmaceuticals in the Environment Pose a Risk to Wildlife?; Society of Environmental Toxicology and Chemistry, SETAC.OnlineLibrary.wiley.com

Marine life all over the world test positive for pharma drugs

Scientists have tested marine life from all over the planet and found the animal tissues contain prescription medications, including antibiotics and synthetic hormones, and drugs for pain, birth control, erectile dysfunction, hair loss, cardiovascular diseases, allergies, acne, weight loss, mood disorders, osteoporosis, and chemotherapy.

Global decline of frogs and other amphibians

It has been known for decades that pharmaceutical drugs, farming chemicals, industrial pollutants, and greenhouse gasses are directly related to the global decline of frogs and other amphibians. Yet, nothing has significant been done about it. With growing population of humans on the planet, more drugs are being taken, and they end up in the environment, and in wildlife – where they cause problems.

Pharmaceuticals and industrial pollutants also are the cause of more and more amphibians being found with both sex organs, with extra limbs, and with other physical defects.

"Frogs appear to be very sensitive to progestofens, a kind of pharmaceutical that is released into the environment. Female tadpoles that swim in water containing a specific progestogen, levonorgestrel, are subject to abnormal ovarian and oviduct development, resulting in adult sterility. This is shown in a study conducted at Uppsala University and published in the journal *Aquatic Toxicology*.

Many of the medicines people consume are released into the environment via sewage systems. Progestorgens are hormone preparations used in contraceptive, cancer treatment, and hormone replacement therapy for menopausal discomfort. Different kinds of progestorgens have been identified in waterways in a number of countries.

Levonorgestrel can cause sterility in female frogs at concentrations not much higher than those measured in the environment.

Female tadpoles that swam in water containing low concentrations of levonorgestrel exhibited a greater proportion of immature ovarian egg cells and lacked oviducts, entailing sterility."
– Sterility in frogs caused by environmental pharmaceutical progestogens, study finds; *Science News*, ScienceDaily.com; Feb. 2011

"Our findings show that pharmaceuticals other than estrogen can cause permanent damage to aquatic animals exposed during early life stages."
– Cecilia Berg, Department of Environmental Toxicology at Uppsala University, 2011

"Amphibians inhabit a wide range of environments, but a common feature is that most species, including terrestrial ones, breed in water. Many amphibians are generalists and spawn in virtually all types of water bodies, including rivers, lakes, ditches, and marine environments, whereas some species are specialized for breeding in, for example, ponds, or rivers. The larval period is a critical period for sex differentiation in many amphibians, and exposure during this life-stage to hormonally active chemicals can disrupt gonadal differentiation. Another sensitive phase of the reproductive cycle is the maturation and release of gametes. During the breeding season, which can last for several months, the adult frogs may be exposed to water-borne chemicals at critical phases of egg and sperm maturation.

Amphibians may be exposed to pharmaceuticals via the water during the breeding period and the larval stages in waterways polluted by municipal wastewater or emissions from drug manufacturing sites, as well as in agricultural areas irrigated with wastewater, or where domestic stock is kept. A number of studies report the presence of hormonally active pharmaceuticals including steroid hormones, aromatase inhibitors, and SSRIs in typical breeding habitats for amphibians, including lakes, rivers, and streams."

– *Risks of hormonally active pharmaceuticals to amphibians: A growing concern regarding progestogens*; National Library of Medicine: National Center for Biotechnology Information; Nov. 2014

As mentioned in the study about progestogens on amphibians, wildlife can be exposed to pharmaceuticals via "agricultural areas irrigated with wastewater." What does that mean? It has to do with "wastewater" from farmed animals, such as the toxic collection "lagoons" of animal waste gathered from factory animal farming with thousands of pigs or cows in enormous steel "barns." All that poo and urine has to go somewhere. It ends up in those lagoons (odd name for something a person might acquaint with a pristine tropical lagoon). Some of it is sprayed onto cropland. What that animal waste contains is all of the pharmaceuticals used on the farmed animals, including what is used to spray on them or "dip" them in to kill parasites, other types of bugs, or inject them with, or feed them to, or otherwise treat them with for illnesses, and to prevent transmittable diseases. By spraying farmed animal waste onto cropland, it exposes the ground, water, air, and wildlife to the pharmaceutical residues. That's one way pharmaceuticals then end up in atmosphere, and in the groundwater, creeks, ponds, rivers, lakes, marshes, and oceans – and us. Most of those crops are feed crops fed to farmed animals. This helps bioaccumulate more toxins in the tissues of the animals. Who eats the meat, dairy, and eggs from factory farmed animals? Humans, and domesticated animals.

Meat, dairy, eggs increase certain human health problems

It should be no surprise that people who eat animal protein-rich diets from factory farms experience diet-related health problems.

"I hope for your help to explore and protect the wild ocean in ways that will restore the health and, in so doing, secure hope for humankind. Health to the ocean means health for us."
– Sylvia Earle

There is the connection between pharmaceuticals and the environment, and wildlife from the mountains to the deep blue seas.

You play a role in the health of the oceans, and wildlife

Consider your role in the health of the oceans.

The oceans start where you live. Even if you live in the middle of a continent, or up on a mountain, water in your area ends up in rivers that lead to oceans.

What you do ends up impacting all varieties of marine life, the animals that feed on the marine life, and the plants and trees absorbing nutrients from those animals.

"How inappropriate to call this planet Earth when it is quite clearly Ocean."
– Arthur C. Clarke

Common household items contribute to water pollution

Most people remain unaware that many of the most common items found in their homes contribute to water pollution. From the soaps and detergents they use – which are often filled with a variety of chemical dyes and scents – to their personal care products, including hair and skin care products and pharmaceutical drugs, and the packaging of those products, they are continually putting substances into their environment that are not good for wildlife.

Fluoride

Toothpaste is one example of a product that not only harms the environment, but also can cause harm to people – even when the product is used as intended. Most toothpaste contains fluoride.

Fluoride toothpaste contains an odorless ingredient called sodium monofluorophosphate, a substance considered for use in chemical weaponry. When reading up on sodium monofluorophosphate, one would have to wonder about its safety, especially when considering the chemical is in products meant for oral care, mouthwash and toothpaste.

There is a connection between the beginning of the fluoridation of drinking water and the use of both industrial farming fertilizers and the uranium used in the atomic weapons program. (For more information, access the Project Censored Award-winning article, *Behind the*

Fluoridation of America: DuPont, the Pentagon, and the A-Bomb, published in the Winter 1997-98 edition of *Earth Island Journal.*)

When reading the labels of toothpaste, you can see that it advises against ingesting the product. The reasoning is that the chemicals are rinsed from the mouth directly after use. But tissues of the mouth absorb chemicals, including drugs.

A required FDA warning on children's toothpaste advises against using an amount of toothpaste larger than a pea for small children, and that if a child swallows toothpaste, they may have to be treated for poisoning. If it is important that children do not eat toothpaste, why is it chemically dyed to resemble candy, flavored to taste like candy, and scented to smell like candy?

Since the 1940s, many municipal drinking water systems have been using a chemical called sodium hydrofluosilic acid and hexafluorosilicate, or fluorosilicic acid, sodium silicofluoride, and sodium fluoride, commonly known as fluoride. Because of this, many who grew up drinking fluoridated water have dental fluorosis stains on their teeth. While the benefit of fluoride is in the direct application to teeth, there is no benefit from drinking it. Consuming fluoride can lead to fluoride damage to the bones, weakening them and making them more susceptible to fracture. Organs harmed by fluoride include the brain, kidneys, and thyroid. Children who drink fluoridated water may also be more susceptible to lead poisoning, especially if they live in homes where lead is in the compound used on water pipe joints. Chemicals that damage human health also damage wildlife. Meanwhile, as most tap water isn't used for teeth brushing, but is used for cleaning and watering of lawns, huge amounts of fluoride and sodium monofluorophosphate are ending up in ground water, in aquifers, in rivers, in lakes, and in the oceans.

Chemicals in cosmetics: cancer, hormone disruptors, etc.

Cosmetics also contain a number of chemicals, including those that cause cancer, mimic hormones, cause birth defects, contaminate water, lead to health problems and birth defects in wildlife, and are simply not good for living things.

Avoid using personal care products containing these substances that are poisonous to you and to wildlife:
Parabens
Monoethanolamine, diethanolamine, and triethanolamine (MEA, DEA, and TEA)
Formaldehyde

Sodium sulfate, amonyl lauryl sulfate, and laureth sulfate
Petroleum jelly
PVA/VA copolymer
Stearolkonium chloride
Synthetic or other coal-tar colors/dyes
Phthalates and synthetic fragrances
Chemical antibacterials
Glyco ether
Phenylenediamine (PPD)

Formaldehyde

One chemical common in many cosmetics, and in hair products, is formaldehyde, which has been identified as increasing the risk of cancers of the breathing passageways and lungs, and of the blood. It is just one of hundreds of chemicals people may be exposing themselves to when they use cosmetics, hair and oral care products, and skin lotions and ointments.

Ingredient list and safety tests not required

The companies making cosmetics can get away with including such chemicals in their products because they are not required to list the ingredients contained in their products.

Cosmetic companies also are not required to conduct safety tests on the ingredients they include in their products.

U.S. bans 11 chemicals. E.U. band over 2,000

While the U.S. Food and Drug Administration has banned only 11 chemicals from use in cosmetics, the European Union has banned over 2,000 chemicals from being used in cosmetics and toys.

"U.S. efforts to promoted consumer safety by regulating chemicals and contaminants in cosmetics are falling further behind the rest of the world.

More than 80 nations – ranging from major industrialized economies like the United Kingdom and Germany, to developing states like Algeria and El Salvador – have enacted rules targeting the ingredients of cosmetics and personal care products. Some of these nations have restricted or completely banned more than 1,600 chemicals from cosmetic products.

By contrast, the U.S. Food and Drug Administration has banned or restricted only nine chemicals for safety reasons.

Federal cosmetics law was enacted in 1938, and has not kept up with the growth of the cosmetics industry. As a result, U.S. regulations of cosmetics products continues to fall short on protecting consumers.

Fortunately, states are not waiting for Congress to act. California and Maryland recently banned 24 of the worst chemicals and contaminants in cosmetics, including formaldehyde, some parabens, and some PFAS."

– On protecting consumers from toxins in cosmetics, U.S. lags at least 80 countries;
Environmental Group; EWG.org; 2021

Chemicals in cosmetic, hair, and skin products end up in Nature

Any products used in cosmetics and in hair, and skin products end up in the rivers, lakes, marshes, and oceans, and in all forms of wildlife.

"Thousands have lived without love, not one without water."
– W.H. Auden

How you live and what you do and purchase matters

The way you live, including the foods you choose to eat, the variety of cleansers you use, the types of medications you take, the forms of transportation you use, and the sources of energy you use, impact all areas of the environment around the planet.

Whenever you eat, consider what types of energy, chemicals, packaging, and other resources and products were used to bring the food to your table.

When you use electricity, consider what types of resources are being used to create the energy – and the pollution that energy production creates.

When you use cleaning products, consider what sorts of chemicals they contain.

When you breathe, consider that the oxygen you depend on is being produced by trees and plants in forests and meadows, and by sea, river, lake, marsh, and water plants thousands of miles away.

"Individually, we are one drop. Together, we are an ocean."
– Ryunosuke Satoro

When you see water, realize you and all life on the planet consist mainly of water. If the water bodies of Earth are polluted, so is wildlife and humanity. If the life in and around the water bodies of Earth dies, so will humanity.

"All of life is interrelated. We are all caught in an inescapable network of mutuality, tied to a single garment of destiny. Whatever affects one directly affects all indirectly."
– Martin Luther King, Jr.

We are reliant on this planet's network of Nature with all of its intricacies of billions upon billions of wild animals, and innumerable plants, bacteria, fungi, and other life forms. But humans have been destroying it.

"The more the planet warms, the greater the impacts. Without rapid and deep reductions in global greenhouse gas emissions from human activities, the risks of accelerating sea level rise, intensifying extreme weather, and other harmful climate impacts will continue to grow. Each additional increment of warming is expected to lead to more damage and greater economic losses compared to previous increments of warming, while the risk of catastrophe or unforeseen consequences increases.

However, this also means that each increment of warming the world avoids – through actions that cut emissions or remove carbon dioxide from the atmosphere – reduces the risks and harmful impacts of climate change. While there are still uncertainties about how the planet will react to rapid warming, the degree to which climate change will continue to worsen is largely in human hands"
– The Fifth National Climate Assessment of the U.S. Government; Nov. 2023

What are you doing to be part of the solution?

This book is one of my ways. It hopefully will inspire and motivate people on every continent to make differences in their lives, and in their communities. They will in turn influence others to do the same.

"What this project is about: the truth about ocean decline. This website seeks to review the current problems, from coral-reef death, to kelp forest over-fishing, to global fisheries depletion. We are not exaggerating the problems – the facts speak loudly enough. We want you to realize how serious the problems have become; learn that today's ocean problems are at the global and ecosystem level."
– ShiftingBaselines.org

No matter where you live, what you do is impacting non-human forms of life, including migrating animals that play a role in nutrient spread, which plays into the health of everything from the oceans to the mountains.

Be part of the solution. Including through a plant-based diet.

Plant and Food System Collapse

Revealing themselves

It is interesting to observe what humans think of as important. A celebrity says something, and within seconds it is splattered across global news and social media. Nonsense is focused on. But important things impacting the lives of everyone, including the plunge in bird and insect populations, and the growing potential for multiple system collapse, are not broadly discussed. What they focus on reveals them. Is it denial? Ignorance? A way to cope? If they ignore it, does it exist?

Most don't know about things greatly impacting them

Ask most people what ocean acidification is. They likely won't know what it is. Ask them where most of the oxygen they breathe comes from, and they likely don't know the answer. Show them the most common food plants, and they likely won't be able to identify many of them. They are even less likely to know how to grow the foods. Tell them a large number of the foods they are used to eating will soon vanish from Earth if the environment continues to degrade, and species continue to vanish, and you might gain their interest, or they will dismiss you as an alarmist, radical, a conspiracist, a doomer, a drag, and unrealistic.

Chances growing of simultaneous global crop losses

"According to Google's news search, the media has run more than 10,000 stories this year about Phillip Schofield, the British television presenter who resigned over an affair with a younger colleague. Google also records a global total of five news stories about a scientific paper published last week showing that the chances of simultaneous crop losses in the world's major growing regions, caused by climate breakdown, appear to have been dangerously underestimated. In media world, a place that should never be confused with the real world, celebrity gossip is thousands of times more important than existential risk.

The new paper explores the impacts on crop production when meanders in the jet stream (Rossby waves) become stuck. Stuck patterns cause extreme weather. To put it crudely, if you live in the Northern Hemisphere and a kink in the jet stream (the band of strong winds a few miles above Earth's surface at mid-latitudes) is stuck to the south of you, your weather is likely to be cold and wet. If it's stuck to the north of you, you're likely to suffer escalating heat and drought."

– George Monbiot, *With our food systems on the verge of collapse, it's the plutocracy v life on Earth*; The Guardian; July, 2023

Using the readily available electronic media, the vast majority of people on the planet can easily research the issues most likely to impact their access to what they need to survive: food, water, oxygen, healthy soil rich in on organisms, a full variety of regional plants, abundant populations of wildlife, and other matters.

Most people do not know what is going on with the environment, or that terrible state it is in, and how it will be impacting their life.

Food, water, and air

What most humans seem completely unknowledgeable of is what is happening to the food plants, the soil, the pollinating animals, the ocean life, and with other factors providing them with food, water, and air.

Bees help keep all life alive

"A growing body of research is revealing just how crucial native bees are as pollinators for many plants. 'They both pollinate our natural systems and – what people don't realize – they are also really important for many of our agricultural crops,' says Scott Black, executive director of the Xerces Society, a nonprofit focused on invertebrate conservation."

– Ula Chrobak, *Why we desperately need wild bees*; Knowable Magazine, Aug. 2023

The *Knowable Magazine* article goes into how wild bees pollinate about one-third of our crops, and the difference between the 20,000 – or so – species of wild bees, and the managed honey bee colonies used by agriculture to pollinate crops. Without wild bees and other wild pollinator animals, the modern-day food system would collapse.

There is an enormous and generally unknown reliance on wild bees to help pollinate not only the crops providing our fruits and vegetables, but also a wide variety of plants that wildlife as a whole depend on for food, habitat, and survival. Without native wild bees and other pollinators, there would be a tremendous reduction in plant and animal life, and that would impact soil creation, soil organisms, water, the weather, the level of oxygen, and the absorption and sequestration of greenhouse gasses. Even most of the flowering plants depend on bees, wasps, butterflies, birds, and other wild pollinators. As far as food, everything from apples to zucchini needs pollinator animals to survive.

Pollinator populations are plunging

The causes behind the loss of bees, hornets, birds, and other wild pollinators include using toxic chemicals on feed crops, lawns, sporting fields, campuses, and other landscaping, using bug sprays and other poisons, clearing millions of acres of land to grow feed crops, and for cattle grazing, and demolishing rainforests and other wildlands. There is also air, water, and land pollution, and the climate change that is causing flooding, drought, and desertification.

Farmland degraded

More and farmland is being degraded through flooding, industrial pollution, toxic chemicals, animal farming, feed crop farming practices, and the spread of desertification that kills off soil organisms – making the soil impossible for plants to take root in... which also leads to fewer wild animals, lower oxygen levels, less food, and hotter weather.

Global warming is stressing plants

One of the issues impacting plant life is the way rising temperatures are stressing the plants, overheating them, and causing their normal way of growing to breakdown. The result is more and more plants are dying from heat, and forests, meadows, and other plantscapes, and the microorganisms dependent on them are increasingly stressed. This includes the plants of the forests and meadows dependent on the forests.

When forests are clearcut, the temperature of the land rises, and native plants can't survive – which reduces native wildlife. This happens all the way down to the soil organisms, including the fungus and bacteria plant life need for root growth and nutrient absorption.

Just as humans host microorganisms – including on and in them, including gut flora – that help them to survive, and play a role in immunity, plants also rely on microorganisms to exist.

Fungus and bacteria needed for plants to grow

"An endophyte, which is predominantly a bacterium or fungus, has an endosymbiotic relationship with the plant host. Endosymbiosis can be defined as a type of symbiosis in which one organism lives inside the other, each benefiting from the relationship.

Although endophytes were found in all studied plant species, an endophyte/host plant relationship is not yet well understood. This may involve competition among endophytic species in the host tissue interposed by production of antifungal metabolites and detoxification of such inhibitors produced by endophytes.

Although mycorrhizal fungi colonize plant roots and reside into the rhizosphere, endophytes live entirely within plant tissues, and may develop within roots, stems, and leaves, sporulate at plant or host-tissue senescence.

Endophytes can cooperate with their host plant by producing secondary metabolites that can protect the plant, providing the ability to defend against predators, helping their hosts to adapt in different stress conditions for survival.

Endophytes encourage plant growth in different ways, such as production of siderophores e.g. enterobactin, and plant growth regulators, such as indole-acetic acid, they can also enhance plant growth through phosphate solubilizing activity. Moreover, endophytic bacteria supply essential vitamins to plants."

– Muna Ali Abdalla and Josphat C. Matasyoh, *Endophytes as Producers of Peptides: An Overview About the Recently Discovered Peptides from Endophytic Microbes*; National Library of Medicine; Sept. 2014

Understanding of interaction with Earth, wildlife, soil, and small life

Somehow, humans have gotten away from the understanding of their interaction with Earth, wildlife, soil, and the smallest of life forms we all depend on for our existence. In ways, humans have become more destructive and completely irrational parasites to the rest of the life forms on the planet.

Symbiosis of microlife

Humans can learn by studying the symbiosis of the microlife of soil, including the smallest of animals living in the soil, the bacteria and fungi, and the root systems of the plants, and how all of it feeds off of decomposing plants, decomposing animal life, the droppings of wild animals, the water system of Nature, and balanced climate gasses.

Nature has a language

The book by Peter Wohlleben, *The Hidden Life of Trees: What they Feel, How They Communicate – Discoveries from A Secret World*, is one that goes into how plants work with the soil organisms and in other ways to communicate and help each other. All plants have ways of interacting with what is around them, especially beneath the surface in their own form of social networking with similar and other life forms.

Trees communicate a sophisticated, silent language

"Trees might be among the lushest metaphors and sensemaking frameworks for knowledge precisely because the richness of what they say is more than metaphorical – they speak a sophisticated, silent language, communicating complex information via smell, taste, and electrical impulses. This fascinating secret world of signals is what German forester Peter Wohlleben explores in *The Hidden Life of Trees*.

Wohlleben chronicles what his own experience of managing a forest in the Eifel mountains of Germany has taught him about the astonishing language of trees and how trailblazing arboreal research from scientists around the world reveals 'the role forests play in making our world the kind of place where we want to live.' As we're only just beginning to understand nonhuman consciousnesses, what emerges from Wohlleben's revelatory reframing of our oldest companions is an invitation to see anew what we have spent eons taking for granted and, in this act of seeing, to care more deeply about these remarkable beings that make life on this planet we call home not only infinitely more pleasurable, but possible at all."
 – Maria Popova, *The Secret Life of Trees: The Astonishing Science of What Trees Feel and How They Communicate*; The Marginalian, the reader-supported blog of Maria Popova: TheMarginalian.org

Trees share nutrients with each other, and create an ecosystem

"Why are trees such social beings? Why do they share food with their own species, and sometimes even go so far as to nourish their competitors? The reasons are the same as for human communities: there are advantages to working together. A tree is not a forest. On its own, a tree cannot establish a consistent local climate. It is at the mercy of wind and weather. But together, many trees create an ecosystem that moderates extremes of heat and cold, stores a great deal of water, and generates a great deal of humidity. And in this protected environment, trees can live to be very old. To get to this point, the community must remain intact no matter what. If every tree were looking out only for itself, then quite a few of them would never reach old age. Regular fatalities would result in many large gaps in the tree canopy, which would make it easier for storms to get inside the forest and uproot more trees. The heat of summer would reach the forest floor and dry it out. Every tree would suffer.

Every tree, therefore, is valuable to the community and worth keeping around for as long as possible. And that is why even sick individuals are supported and nourished until they recover. Next time, perhaps it will be the other way round, and the supporting tree might be the one in need of assistance."
– Peter Wohlleben, in his book *The Hidden Life of Trees*

World is becoming less green and more dryer, and dead

"A 2010 study in *Science* was among the first to demonstrate that the greening increases of the 1990s had stalled or reversed. That study also suggested that the declines were probably water-related.

That's not to say every last corner of Earth is losing vegetation. Some recent studies have revealed that parts of the Arctic are 'greening' as the chilly landscape warms. And there's increasing plant growth still happening in other regions of the world, as well.

But on a global scale, averaged across the entire planet, the trend is pointing downward.

The declines challenge an argument often presented by skeptics of mainstream climate science to downplay the consequences of global warming: the idea that plants will grow faster with larger amounts of carbon dioxide. The argument hinges on the idea that food supplies will increase.

It's largely a red herring, as climate scientists have patiently explained for years. Rising CO_2 does benefit plants, at least up to a point, but it's just one factor. Plants are also affected by many other symptoms of climate change, including rising temperatures, changing weather patterns, shifts in water availability, and so on."
– Chelsea Harvey, *Earth Stopped Getting Greener 20 Years Ago*; E&E News, *Scientific American*; Aug. 2019

Add damaged, decreasing, and vanishing soil organisms to the list of causes of the decrease in Earth's vegetation.

No idea what they are damaging, and sending into extinction

With so much damage being done to Earth and its forests and other plant and animal life, and so much of it remaining unexplored, humans have no idea what they are damaging, what they might be sending into extinction, and how those various forms of life lost forever may have played roles in future cures and other solutions – if only to have made existence better for all other life forms, and possibly even possible.

Animals and most organisms can't exist without plants, and plants can't exist without animals and a healthy variety of organisms – especially soil organisms that both help transfer nutrients into plant roots and tissues, and produce nutrients for plants to survive – including plants all animals need for survival.

Trees are having problems in the increasing global heat

"Global warming is driving leafy tropical canopies close to temperatures where they can no longer transform sunlight and CO_2 into energy, threatening total collapse if the thermometer keeps climbing, according to a study Thursday."
– Raul Arboleda, *Tropical forests nearing critical temperatures*; france24.com, Aug. 2023

NASA has an ECOSTRESS system (ECOsystem Spaceborne Thermal Radiometer Experiment on Space Station) mounted on the International Space Station that can measure forest canopy temperatures. The system has revealed that an increasing number of trees are being impacted by rising temperatures.

More and more trees are dying in the Amazon simply because of the rising heat. A loss of native wildlife also stresses or damages them, as does the increasingly dry land lacking in the microorganisms that help them to grow. More millions of the trees have been cut down to make room for cattle grazing, for monocultured feed crops, for sun-grown coffee, for oil palm plantations, and other crops.

Fragmentation of the tropical forests, a sin of humanity

A quick Google satellite image search of the Amazon reveals how fragmented the forest has become.

Fragmenting the forests increases the heat as trees keep air cooler than shorter plants do. Thick forest canopies keep the ground many degrees cooler than what is at the tree tops. Even the process of photosynthesis in leaves cools the air. Combined, trillions upon trillions of leaves also play into the water, carbon, and all life cycles all over the planet – from the mountains to the bottoms of the seas.

Humans are killing the rainforests, which consist of nearly half of the forests on the planet, and host most of the plant species on Earth.

The fewer trees that exist, the fewer plants there are, the less greenhouse gasses get absorbed, and the hotter the planet becomes.

Fewer forests means fewer animals means weakened Nature

The fewer trees and other plants there are, the less of a chance the animals can survive.

Animals help to fertilized and pollinate the plants. With fewer animals the less chance both animals and plants – and soil organisms – have of surviving. And the hotter Earth gets.

Humans: the most destructive being = the ecocide species

Life supports life. Except human life seems to do the opposite. Humans continue to prove they are the most destructive being.

The human-caused climate breakdown is ecocide. It is extermination of wildlife. It is leading to what is killing more and more thousands of people around the world through climate breakdown, heat, food shortages, storms, and other "natural disasters" that are actually caused by humans.

Scorched leaves, damaged trees, weakening structures, and collapse

Leaves among the upper branches of trees can heat up far faster than other areas of the trees. This can scorch the leaves, and lead to a weakening of the entire tree structure, leading to collapse.

"You heat the air by two to three degrees, and the actual upper temperature of these leaves goes up by eight degrees.

The Amazon is currently experiencing higher levels of mortality than Central Africa, and that could possibly be due to the high temperatures we've seen there."
– Christopher Doughty, Northern Arizona University

"Even when the surrounding air is significantly cooler than the leaves themselves, some .01 percent of individual leaves can reach a critical temperature at which the enzymes required for photosynthesis go through a process called denaturation."
– Meghan Bartels, Tropical forests may be getting too hot for photosynthesis; Scientific American, Aug. 2023

"It is getting too hot for trees, and for tropical rainforests."
– Stephanie Pau, Florida State University

Tropical forests are necessary to all Earth life

Tropical forests are necessary for not only absorbing and sequestering greenhouse gasses, and producing oxygen, but it is where many of the food plants humans rely on originated. Among the foods from the rainforests are varieties of fruits, berries, vegetables, and grains.

Everything from acai, avocadoes to chocolate, coconuts, coffee, corn, ginger, black pepper, pineapples, potatoes, hot peppers, vanilla, and yams originated in the rainforests.

Most common human foods and medicines are from rainforests

Scientists estimate that at least 80% of the currently common foods humans eat originated in the rainforests. Many herbs, spices, and medicines also are from the rainforests. As is much of every breath of oxygen you take in.

The water you drink is partially because of rainforests

Because the rainforests play such a role in the water systems and rain storms of Earth, you may as well count the water you drink as partially a product of distant rainforests.

We won't know the foods & medicines humans sent into extinction

With so many varieties of plants going extinct, we don't know and will never know what sorts of medicines and foods will not be available – as they hadn't been discovered by the time the plants went extinct.

Dire situation facing global food supplies

What is going on with the rainforests is only part of the dire situation facing the global food supplies.

The summer of 2023 had a wide variety of awful things happen in relation to climate breakdown, and to the food system humans are reliant on. Those things include thousands of acres of farmland being flooded with polluted water in California (making that farmland too polluted to use for years), thousands of acres of crops in Italy dying in the heat, cropland in Germany being destroyed by fierce rain storms, India blocking the export of certain crops, leading to a global rice shortage (because the combination of heat and flooding has greatly reduced the harvests), and so forth and so on it went on every continent.

Additionally, intense environmental catastrophes of the summer of 2023 included hundreds of fires burning everywhere from Greece and other European countries, to Canada, to the extent that air quality for thousands of miles had been degraded. There was also the incineration of most of the town of Lahaina, Maui, Hawaii, where the number of people who perished in the fire might never be known. During the same month, the ocean surface temperatures in many areas across the globe set new records as the warmest ever recorded.

It seems likely that the previous paragraph describes what will become normal, with the possibility the we will experience far worse.

Melting bee hives, and bees that won't forage if it is too hot

One problem with global warming is that it can melt bee hives. It is one of the things the increasing heat has been doing in the American Southwest. It happens when temperatures both get too hot and bees also lack access to water. In 2023, Arizona had experienced many weeks of temperatures of at least 110 degrees, causing hives to melt. When a hive melts, it sends bees scouring for food, even raiding other hives. The bees can end up fighting to death as there is a blockade from allowing bees from other colonies inside, and also to limit the number of bees in a hive so that it doesn't raise the heat. In extreme heat, as a protective measure, bees will even block those from the same hive from reentering.

If temperatures get too hot, bees simply won't forage for food. Without food, they can't survive. Without bees foraging, they don't pollinate plants, including food crops. This is another layer of problems created by climate ruin.

Shortage of rented bee hives to pollinate crops

With so many farms relying on bees to pollinate crops, there also has been a shortage of rented hives. These are hives trucked from farm to farm to pollinate crops. The heat damaging hives has magnified the problem. This is part of what is causing some crop failure, as is the dying off of native wild bees. These situations apparently will become more common on a heating Earth with radical weather changes.

Global food impacted by heat, drought, floods, degraded soil, a plunge in pollinator populations, as well as increased pollution, greed, and wars

A person can research the conditions on each continent, and learn about many concerning issues relating to what is going on with the land, the water, the pollinators, and the food plants of the world.

As so many varieties of food crops around the planet are being impacted by heat, drought, floods, and soil degradation, food shortages and higher food prices are things more people are dealing with. Whether they like it or not, they will get used to it.

Climate breakdown impacting everyone

Presently, because of the industrialized food system, in one way or another, climate breakdown conditions are impacting everyone.

Most people are reliant on purchased food for every calorie

There are few people who are completely reliant on what they grow or gather. The vast majority of people on the planet have become reliant on food grown in distant regions, which plays into nearly every calorie they eat being purchased. The situation plays into chronic hunger, rising food prices, soil degradation, and wildlife loss, which are all increasing.

Food security is uncertain under future climate change: collapse

"Food security is uncertain under future climate change, but is there a threat of food system collapse? Now research assesses the probability of weather hazards occurring at the same time in the world's major breadbaskets and reveals that the weather-related component of this risk could be increasing.

Over the short term, crop failure can lead to price spikes, migration, local conflicts, and famine. In the longer term, increases in the frequency of these failures could reduce the adaptive capacity of society to respond to future shocks, exacerbating the short-term impacts."
– Zia Mehrabi, *Food System Collapse*; Nature.com; Oct. 2020

Combination of climate and food system breakdown: deep trouble

The combination of climate and food system breakdown is becoming more troublesome and common. Human and other animal life is in deep trouble. At what point is it going to become enough for humans to do enough about it to turn this collapse around? Or, will they? It seems likely that millions of people will die every year from food insecurity, heat, water shortages, and other matters relating to the changing situation of Nature caused by human activities.

Has the tipping point passed = is it now too late?

Has the tipping point been passed?
Have too many species already gone extinct?
Are there already too much greenhouse gasses in the atmosphere for humans to be able to do enough about it to protect the future of wildlife?

World leaders cater to corporations, the greed class, and religions

The current selection of "world leaders" and their governments cater to the corporations, and continue to allow the fossil fuels, animal farming, plastics, and motor vehicles industries to destroy Nature and Earth. As they make choices based on religion and racism.

The ultra-rich: egos, narcissism, selfishness, greed & delusion

"What the ultra-rich want is to sustain and extend the economic system that put them where they are. The more they have to lose, the more creative their strategies become. As well as the traditional approach of buying media outlets and pouring money into the political partis that favor them, they devise new ways of protecting their interests.

Corporations and oligarchs with massive fortunes can hire as many junktanks (so-called thinktanks), troll farms, marketing gurus, psychologists, and micro-targeters as they need to devise justifications and to demonize, demoralize, abuse, and threaten people trying to sustain a habitable planet. The junktanks devise new laws to stifle protest, implemented by politicians funded by the same plutocratic class."

– George Monbiot, *With our food systems on the verge of collapse, it's the plutocrats v live on Earth*; TheGuardian.com, Aug. 2023

We are heading towards 3 to 4 degrees warming this century

"If we think about where we are heading, let's be clear: We are over 30 years, 32 years since the first major scientific report on climate change that came out in 1990. And so, I think when we judge where we are heading, we have to say, 'Well, what have we done since 1990?'

We've watched emissions go up, year after year after year. They are now over 60% higher, per year, than they were in 1990. So, you will hear lots of rhetoric, lots of good words, lots of optimism about the future. But given we've known about this subject – and apparently been working on it for 30 years – the trend line tells us we are heading towards 3 to 4 degrees centigrade of warming across this century. An absolute climate catastrophe. And it's a catastrophe for all species, including our own. And so that is the direction of travel.

Now, that direction of travel does not have to continue. But the current trend line tells us that all we are doing so far is giving rhetoric and optimism and greenwash, and not driving the levels of change that are necessary to stay within the 1.5 to 2 degrees framing of the Paris agreement.

When we think about 3 or 4 degrees centigrade, let's be clear: We have no historical precedent in human history of these sorts of temperature changes. And they are occurring overnight. And they don't just occur over this century.

Firstly, we know that things like sea level rise will keep going for hundreds of years after that, and that we are locking in – absolutely locking in – really high levels of sea level rise, maybe 7 to 8 more meters. So we may only across this century see 1 or 2 meters – which will be devastating for many of our coastal cities. And, of course, most of the population of the world lives near the coast. So that would be devastating for our existing communities. But we are locking in this devastation for centuries to come. But we're also changing very significantly how we will produce our food. Whether we produce enough food, where our food will be produced. And that's because we're changing the complete weather patterns of our Earth. We are changing rainfall patterns. We're changing insect pollination of our crops.

So all of this plays out one disaster after another. So any single one of them we might think, 'Oh, we can resolve, we can deal with that.' But when you bring all of these together, occurring almost overnight, you're talking about occurring almost overnight, you're talking about the collapse of our modern society. You're talking about the collapse of most of our sort of emblematic ecosystems.

So, this is not a future that we should in any way be heading towards. And we should be doing everything we can to avoid it.

The sad fate of affairs is thought that we're doing nothing to avoid it. There's plenty of talk, but no action. And what we have to bear in mind is that climate only responds to action.

The physics respond to how much carbon dioxide and other greenhouse gasses we put in the atmosphere.

We can talk about efficiency. We can talk about green growth, and all of this stuff. It's meaningless. What really matters is keeping the emissions out of the atmosphere.

What we've seen in the last few years is that countries have made these various promises about what they are going to do about climate change. And the really optimistic reading of that, often held, often given by people like the U.N. and some other climate academics, they say, 'Oh, we're heading towards 2.7 degrees of warming.'

Well, first, let's be clear. 2.7 degrees of warming at a global average level is itself a disaster. I mean, these temperatures are not spread evenly around the planet. So that is like moving to a different planet. It's not the one we live on today.

But, in addition to that, holding to 2.7 relies on a huge amount of optimism by the analysts, and by the U.N. It's not embedded in the commitments the governments are making.

You look at those commitments, and you extend them out beyond the time frame that they're normally talking about, which is only to say to 2030, and then we are much more likely to be heading toward 3 or 4 degrees centigrade of warming.

This 2.7 is optimistic, but because it is relying on future generations, our children, and our grandchildren to develop technologies that will remove carbon dioxide from the atmosphere. Not small quantities of carbon dioxide. Hundreds of billions of tons of carbon dioxide to be removed from the atmosphere, and stored safely somewhere underground.

That we are relying on these future generations to develop the technologies that we don't have today. We have a few pilot schemes, and a few ideas, and a few professors' minds. There is nothing out there of any scale that we're talking about. So we are in time to imagine a future that actually removes carbon dioxide from the atmosphere because we are running too scared of the political repercussions of actually driving the emissions out of the system today.

Engineering can do a huge amount of things, but it cannot perform miracles. And yet a lot of models that we have, what we have done in those is in bed with all sort of sometimes real technologies, often pseudo technologies, and then we hide behind them. They are a façade to avoid asking the difficult political and equity based questions. And you need to really guard against that.

When people are telling you, 'Oh, we've got to 2050 to make these changes,' and even then when they say 2050 and net zero, net zero is a real dangerous turn in my view. And if you hear the language of net zero, I'd be very cautious about the optimism of the person who's saying it actually has. Unpick it, reveal what's behind it, and you'll realize what they mean is not zero emissions, not net zero, not zero emissions. I always call it not zero is Latin for 'kick the can down the road.'

You need to look at: what are the technologies hiding, and what they are often hiding is this deep inequality in emissions.

Who are the people presiding over most of the emissions and the way that they are avoiding those questions is to say we can do it with this technology or that technology in 2030, in 2040, in 2050. And of course, well beyond that, because a lot of these net-zero models are assuming technologies in 2070 and 2090, 2100. And these are the technologies that don't really exist today.

Many of us who work in the climate change realm, I think, are hiding behind this because we've done very, well, out-the-system thinking, and we have nice places to live. We have the benefit of travel, easy travel. We can afford the fuel. We're not in a cost of living crisis, like many people are facing around the world today. So, for us, life is good, life is quite rosy. And we don't like to see ourselves as part of the problem. And one of the ways around that is to delude ourselves. And in doing that, other people as well, that actually technology is the savior.

Technology is part of the picture, is a prerequisite. But it needs to go hand in hand with fundamental, profound social change by those of us who are responsible for the lion's share of the emissions.

And I have to be honest and say that my judgment, my best guess, as someone who has worked in this for years, is that we are going to fail. We're going to go to 3 or 4 degrees centigrade of warming, and we'll have to live through – or die from – all of the repercussions that we will have.

That is a terrible prospect, and one that I think we have to try everything we can do to avoid.

But the message of hope – if there's any thread of hope in this – is that it is a choice to fail.

We have so far repeatedly. I'm going to say 'we.' What I mean is, effectively our leaders politically, academically, in the journalistic community, across the board, those people who frame this debate have chosen actively to fail for three decades. And when they have breakfast with their own children, then I hope they are thinking about what they have deliberately – what we have deliberately – imposed upon their future.

But, as I say, it's a choice. And the great thing about choice is we can choose a different way out of this.

Now, whether we can still hold to 1.5 degrees, it looks incredibly unlikely to me. But incredibly unlikely doesn't mean to say that it's impossible. It is only impossible if we don't try.

And so the real trick here is to try. To try really hard.

What examples do we have of rapid change? Because we don't have any examples of this.

History is littered with rapid change. From time to time these things occur. Particularly when we see there's a collective agreement that we're in a certain situation.

So with COVID now, obviously a deep tragedy, but what we saw was a global response. Now there are lots of responses where we're not particularly in the right direction, and in an emergency that's often going to be the case.

Nevertheless, we did see a global response to COVID.

We saw a global response to the banking crisis back in 2007-2008. In my view, not in a healthy, helpful, sustainable way. But nevertheless we saw a global response.

We have people like Roosevelt's fireside speeches going back to the 1930s that really were radical changes that were being proposed to the social norms of the time.

We've been through the suffragette movement.

We've been through all of the changes in race law, and so forth across our history.

We need to take those examples and accelerate them, and say, 'Yes, we can drive change very rapidly.'

It is a choice to fail, and it is a choice to succeed. And if we sit back and wait for the great and good to deliver this change, then we will fail.

It does come down to all of us to play our role as best that we can in doing this.

And thankfully we are seeing early signs of this with some of the civil society movements who are really working very hard to try and change the agenda, and change the dialogue, and the mood music over the last two or three years – maybe the last five years now – has come from that group. It hasn't come from the professors or the academics. It's come from civil society. And to me that's really where the start and nugget of hope arises."
    – Kevin Anderson, climate scientists and professor of energy and climate change, University of Manchester

"Global food traditions have incorporated a wide array of agricultural species and varieties throughout history and around the world, and agricultural biodiversity is also being destroyed in the Sixth Great Extinction. Our growing population creates pressure for increased food production. This has led to more intensive single-crop production, or monoculture. These combined factors have resulted in the loss of profound agricultural biodiversity.

The FAO estimates that we have lost 75 percent of our agricultural variation during the twentieth century. The result of this loss in agricultural biodiversity is a potential crisis in food security, as diverse varieties of systems provide a safeguard from future threats, adversity and ecological changes.

The destruction of our agricultural biodiversity could lead to a more expansive version of the Irish potato famine, which resulted in a million human deaths because of a lack of diversity in food crops.

As human activity continues to change the Earth's climate, it is uncertain which characteristics will be required in our future crops, but they may be nothing like those of the increasingly few crops we have chosen to mass produce now."
    – The Sixth Great Extinction, VoicesForBiodiversity.com

To safeguard the air, water, soil, climate, and wildlife, it is important to safeguard all plants native to each region of Earth. This includes those edible for humans, and those edible for local wildlife.

Restoring Nature:
Soil, Water, Plants, and Wildlife and their Terrain

Climate crisis and global conflict

"We are at the beginning of a *MadMax* era where the fight over basic amenities will lead to globalized conflict, just as these amenities are diminished by the climate crisis."
— George Tsakraklides, biologist and author of *In The Grip of Necrocapitalism: The Making and Breaking of A Psychonomy*

Be part of the restorative revolution

"Our ancestors fought through genocide. They fought through that trauma. And not only did they survive that trauma, they passed down knowledge that built the societies that were built on today. So this knowledge that has been passed down for thousands of years can be accessed and applied to your daily life, no matter where you are, no matter where you are from, because the industrial revolution is over. Now, if we want to carry on life on Earth, we need to be part of the restorative revolution. Whatever that looks like to you, just make sure you get your hours in."
— Samuel Gensaw, Yurok Tribe of Northern California. Ancestral Guard. NatureRightsCouncil.org, NativeFoodSystems.org

"It's this movement among all indigenous people, that they're finally, they're listening. They are feeling this. They hear it. It's like music, when you hear the drum, it's calling you. I feel that from all different cultures. They hear the same drum. It's Mother Earth.

305

Mother Earth's heart is beating, and she's talking to all of us, that we need to do something."
– Margie May Bendle Stevens, San Carlos Apache Tribe

Protect rivers, lakes, oceans, and all life dependent on them:

1) Drastically reduce your use of fossil fuels, including petroleum, coal, and natural gas. This includes the use of fossil fuel plastics, or otherwise non-compostable plastics.

2) Choose to eat organically grown foods.

Work to distance yourself from depending on foods grown using synthetic chemical fertilizers, pesticides, herbicides, insecticides, miticides, and fungicides, and sold in toxic packaging.

Support local organic produce farmers in your region.

Where possible, grow some of your own food using organic gardening methods.

Compost your organic food scraps to build healthy soil rich with microorganisms that work in combination with plant roots to grow healthy plants rich with endophytes. The microorganisms need the plants, and the plants need the microorganisms.

Benefits of a plant-based diet

"Shifting to plant-based eating habits can reduce the greenhouse gas emissions of our food system by 49%, reduce food's land use by 76%, and reduce freshwater use by 19%. And it's heartening to know that everyone can do their part by committing to eat just a few more plant-based meals per week.

Today's unsustainable global food production methods and consumption habits are responsible for about one-third of all human-caused greenhouse gas emissions, nearly double that of plant-based foods. Emissions from the livestock sector account for as great a share of global greenhouse gasses as the exhaust from all cars, trucks, motorcycles, airplanes, trains, and boats around the world.

The existing food system also puts enormous pressure on agricultural land and water. Meat, dairy, and aquaculture production systems use the vast majority (83%) of the world's farmland – despite providing just 18% of global calories, and 37% of protein. Farmed animal production has also dominated land-used change, pushing crop production and pastures into wild habitats and contributing to an alarming rate of species extinction."
– Kitty Block, *More animals than ever before – 92.2 billion – are used and killed each year for food*; Blog.HumaneSociety.org; June 2023

3) Eat low on the food chain:

Don't eat meat (including fish and birds), dairy, or foods containing them.

Follow a plant-based diet free of junk, dyes, clarified sugars, and highly-heated extracted food oils (fried and sauteed).

"You will save 890 gallons of water equal to 22.5 bathtubs for each day you eat a plant-based diet. You save 148 square feet of rainforest and 1.7 square feet of topsoil for every day you eat a plant-based diet. By following a plant-based diet once a week, it is equivalent to driving 1,000 miles less every year."
– University of California Santa Barbara Green Monday Campaign

4) Compost your kitchen scraps.

Don't send them to landfills, trash dumps, or incinerators.

Get local restaurants, cafes, and food production facilities to do the same.

Some cities and towns are now requiring people to compost food scraps. Don't wait for laws to tell you.

5) Use household products, including plant-based cleansers that are not only biodegradable, but compostable.

Avoid using chlorine bleach, or cleaning products containing petroleum or coal extracts.

(Search: How to make safe household cleansers.)

6) Reduce your use of plastic.

Look for compostable plastics consisting of 100% plant substances. That means compostable plastics made of corn, sugar cane, wheat, potato, hemp, soy, bamboo, or other plants.

(Search: Compostable hemp plastics.)
PlasticPollutionCoalition.com
PlasticFreeTimes.com.
MyPlasticFreeLife.com
PostCarbon.org

7) When you shop:

Use a nonsynthetic cloth bag, not paper or plastic.

People think tree paper is safe and plastic isn't. The truth is, for a number of reasons, including because of the amount of resources used in creating them, neither tree paper nor plastic bags are good for the environment.

In 2022, Americans used over ten billion paper grocery bags, and about 100 billion plastic bags.

Those plastic bags end up as trash of some sort every day, including on the land, and in rivers, lakes, marshes, and oceans.

Paper bags are most often made from trees. We shouldn't be cutting down the forests to create such "disposable," single-use products.

The system needs to be converted to only use paper bags and cardboard made of locally-grown hemp.

Plastic bags are made from fossil fuels, and we are well aware of how much environmental damage petroleum and natural gas have done and are doing to the planet, including in the drilling, fracking, and extraction, the shipping and processing, and in the burning of it.

Please use nonsynthetic cloth bags.

8) Get local stores and businesses to reduce and eliminate their use of single-use plastics.

9) Plant and protect trees. Restore and protect forests. Restore and protect wildlife habitat. Rewild Earth. Support organizations doing the same – especially in your region of Earth.

(Search: Forest restoration organizations.)

10) Legalize industrial hemp farming.

Hemp and bamboo can provide the materials we commonly depend on trees to provide, including paper, cardboard, and lumber.

Unlike trees, in addition to flooring and plywood, hemp and bamboo can also be made into fabric, food, and engine fuel.

An acre of either hemp or bamboo absorbs more greenhouse gasses than an acre of trees. This helps sequester the greenhouse gasses to reverse global warming, ocean acidification, and other climate breakdown.

Unlike cotton, which is the most fertilized and pesticide-sprayed and water-hungry crop, hemp and bamboo grow easily without fertilizer while providing material that creates fabric that is stronger, softer, and lasts longer than cotton.

Visit: VoteHemp.com and the Hemp Industries Association site: TheHIA.org.

11) Reduce pollution.

Reduce your use of products that pollute, and that fill trash dumps. Reuse, recycle, repurpose. Buy less. Consider second-hand items.

12) Pedal instead of driving.

Support and promote bike culture.

If you live in a city or large town, there is likely a bike repair and bike building workshop, and biker social groups.

13) Shop less.

Stop trying to replicate corporate imagery in your life.

Avoid mimicking commercial imagery and celebrity culture.

Stop going into debt to purchase things you don't need, which gather next to other things you bought that you didn't need.

"Mindless, mesmerized consumatrons, humans will always be preoccupied with their latest toys, even as the world burns and floods. Our blind trust in technology was the biggest mistake in our evolution."
– George Tsakraklides, biologist and author of *In The Grip of Necrocapitalism: The Making and Breaking of A Psychonomy*

"History will not only judge us by our mistakes, but by what we do to fix them."
– Jacques Yves Cousteau

Documentaries to watch:
• *Acid Test*, by the Natural Resources Defense Council
• *The Big Fix*
• *Blue Planet: Ocean World: Frozen Seas*
• *Breaking Boundaries: The Science of Our Planet*
• *The Cove*
• *Cowspiracy*
• *The End of the Line*
• *Forks Over Knives*
• *Gasshole*
• *Gather*
• *An Inconvenient Truth*
• *The Need to Grow*
• *Seaspiracy*
• *Tipping Point*, about ocean acidification
• *The Whale Warrior: Pirate for the Sea*

Watch documentaries about trees and the rainforests.

Watch documentaries about the oceans.

There are all sorts of documentaries about Nature.

Watch documentaries about trees, forests, wetlands, marshes, lakes, rivers, and ponds.

Watch documentaries about environmental issues, permaculture, co-op living, green living, organic gardening, alternative fuels, and the protection of wildlife.

World scientists warn

"Excessive extraction of materials and overexploitation of ecosystems, driven by economic growth, must be quickly curtailed to maintain long-term sustainability of the biosphere. We need a carbon-free economy that explicitly addresses human dependence on the biosphere and policies that guide economic decisions accordingly. Our goals need to shift from a GDP growth and the pursuit of affluence toward sustaining ecosystems and improving human well-being by prioritizing basic needs and reducing inequality.

Still increasing by roughly 80 million people per year, or more than 200,000 per day, the world population must be stabilized – and, ideally, gradually reduced – within a framework that ensures social integrity. There are proven and effective policies that strengthen human rights while lowering fertility rates and lessening the impacts of population growth on greenhouse gas emissions and biodiversity loss. These policies make family-planning services available to all people, remove barriers to their access, and achieve full gender equality, including primary and secondary education as a global norm for all, especially girls and young women."

– World Scientists Warning of a Climate Emergency, a 2019 report by the Alliance of World Scientists, signed by 15,000 scientists; *BioScience;* updated Oct. 2023

This site contains information about modern-day extinction:
MassExtinction.net

These organizations work to protect marine and forest life:
Algalita, Algalita.org
Ancient Trees, AncientTrees.org
Common Vision, CommonVision.org
EarthEcho.org
Earth First, EarthFirst.org
Endangered Species International, Endangeredspeciesinternational.org
Environmental Justice Coalition for Water, EJCW.org
Fishing Hurts, FishingHurts.com
Forest Advocates, ForestAdvocates.org
Forest Council, ForestCouncil.org
Forest Ethics, ForestEthics.org
Forests Forever, ForestsForever.org
Forest Protection Portal, Forests.org
GreenPeace, GreenPeace.org
The Greywater Guerillas, SFUAS.org
Harp Seals, HarpSeals.org
International Rivers, InternationalRivers.org
Leatherback Trust, LeatherBack.org

Native Forest, NativeForest.org
Natural Resources Defense Council, NRDC.org
The Ocean Project, TheOceanProject.org
Ocean Protection Coalition; oceanprotection.org
Protect Our Woodland, ProtectOurWoodland.co.uk
Rainforest Action Network, RAN.org
Rainforest Web, RainForestWeb.org
Rainwater Harvesting, RainWaterHarvesting.org
Redwoods Rising, SaveTheRedwoods.org
Reef Resilience, ReefResilience.org
Rising Tide Australia, RisingTide.org.au
Rising Tide North America, RisingTideNorthAmerica.org
Rising Tide UK, RisingTide.org.uk
Sanctuary Forest, SanctuaryForest.org
Save Oaks, SaveOaks.com
Save the Manatee, SaveTheManatee.org
Save the Redwoods League, SaveTheRedwoods.org
Sea Otters, SeaOtters.org
Sea Turtle, SeaTurtle.org
Sea Shepherd, SeaShepherd.org
Sequoia Forest Keeper, SequoiaForestKeeper.org
Tas Forests, TasForests.green.net.au
Tree Musketeers, TreeMusketeers.org
Tree People, TreePeople.org
Trees for Life, TreesForLife.org
Trees for the Future, TreesFTF.org
Trees Foundation, TreesFoundation.org
Water Keeper Alliance, WaterKeeper.org
We Save Trees, WeSaveTrees.org
The Wilderness Society, Wilderness.org
Wildlands CPR, WildlandsCPR.org
World Watch, WorldWatch.org
World Water Council, WorldWaterCouncil.org

We're the last surviving hominid, and not doing well

"We're the last surviving species of hominid, and haven't done well so far, and now we're about to take down hundreds of thousands of species with us. Well, we've created this mythology where everything revolves around us, that we're the center of creation and it doesn't really matter what we do to the rest of the world. The other problem of course is that a large percentage of humanity is living this pie-in-the-sky fantasy where they believe that real life begins when they die and they go to some afterlife somewhere – so that the planet is expendable. In fact, I've had some Christians come to me and say, 'You know, you're interfering with God's plan,' that, 'God has ordained that these

311

resources are to be used while we're here and Jesus is going to be upset if we don't use the resources that he's provided for us.'"

– Paul Watson, in the documentary, *The Whale Warrior: Pirate for the Sea*; PirateForTheSea.com

Bike Culture

The world must quickly implement efficiency and conservation

"The world must quickly implement massive energy efficiency and conservation practices, and must replace fossil fuels with low-carbon renewables, and other cleaner sources of energy safe for people and the environment. We should leave remaining stocks of fossil fuels in the ground, and should carefully pursue effective negative emissions using technology, such as carbon extraction from the source and capture from the air, and especially by enhancing natural systems. Wealthier countries need to support poorer nations in transitioning away from fossil fuels. We must swiftly eliminate subsidies for fossil fuels, and use effective and fair policies for steadily escalating carbon prices to restrain their use."

– *World Scientists Warning of a Climate Emergency*, a 2019 report by the Alliance of World Scientists, signed by 15,000 scientists; *BioScience*; updated Oct. 2023

One of the easy ways many people can reduce their fossil fuel use is by biking, instead of driving.

"Get a bicycle. You will not regret it."
– Mark Twain

"Every time I see an adult on a bicycle, I no longer despair for the future of the human race."
– H. G. Wells

"A human body and a bicycle are the perfect synthesis of body and machine."
– Richard Ballantine

"It is curious that with the advent of the automobile and the airplane, the bicycle is still with us. Perhaps people like the world they can see from a bike, or the air they breathe when they're out on a bike. Or they like the bicycle's simplicity and the precision with which it is made. Or because they like the feeling of being able to hurtle through air one minute, and saunter through a park the next, without leaving behind clouds of choking exhaust, without leaving behind so much as a footstep.
– Gurdon S. Leete

"The bicycle is the most civilized conveyance known to man. Other forms of transport grow daily more nightmarish. Only the bicycle remains pure in heart."
– Iris Murdoch

"Let me tell you what I think of bicycling. I think it has done more to emancipate woman than anything else in the world. It gives women a feeling of freedom and self-reliance. I stand and rejoice every time I see a woman ride by on a wheel. The picture of free, untrammeled womanhood."
– Susan B. Anthony

"The bicycle is just as good company as most husbands and, when it gets old and shabby, a woman can dispose of it and get a new one without shocking the entire community."
– Ann Strong

"It is by riding a bicycle that you learn the contours of a country best, since you have to sweat up the hills and coast down them. Thus you remember them as they actually are, while in a motor car only a high hill impresses you, and you have no such accurate remembrance of country you have driven through as you gain by riding a bicycle."
– Ernest Hemmingway

"The bicycle had, and still has, a humane, almost classical moderation in the kind of pleasure it offers. It is the kind of machine that a Hellenistic Greek might have invented and ridden. It does no violence to our normal reactions: It does not pretend to free us from our normal environment."
– J.B. Jackson

"Nothing compares to the simple pleasure of a bike ride."
– John F. Kennedy

If you ride a bike, please **have lights and reflectors on your bike**, including a rear red light, and a front white light. In addition to lights, install side, front, and rear reflectors. And, **wear a helmet**. Also, simple **biker gloves** will protect the hands. Obey traffic laws. Don't wear headphones that block sounds you need to hear for safety while riding. Put a bell on your handlebars. Ringing the bell is an easy way to let people know you are approaching. **Ride safely**.

"Think of bicycles as rideable art that can just about save the world."
– Grant Peterson

Books:

• *Cycling for Profit: How to make a living with your bike*, by Jim Gregory
• *Pedaling Revolution: How cyclists are changing American cities*, by Jeff Mapes

Bamboosero, bamboosero.com. Bamboo bikes made in Africa.
Bicycle Civil Liberties Union, BCLU.org
Bicycle Paper, BicyclePaper.com. Seattle, Washington
Bicycle Safe, BicycleSafe.com
Bicycle Transportation Alliance, BTA4Bikes.org
Bicycling Info, BicyclingInfo.org/pp/exemplary.htm
Bicycology, Bicycology.org.uk
Bike Blips, Bikeblips.dailyreader.com
Bike Boom: BikeBoom.com
Bike Cart, bikecart.pedalpeople.com
Bike 4 Peace, Bike4Peace.org
BikeRoWave, BikeRoWave.org
Bike School, BikeSchool.com
Bikes at Work, 129 Washington Ave., Ames, IA 50010; bikesatwork.com. Sells bike trailers that can be used for hauling or transporting. They also provide information on starting a bike-based business: bikesatwork.com/cycling-for-profit
Bikes Not Bombs, BikesNotBombs.org
Bike Trailer Shop, biketrailershop.com
Bike Trax, Biketrax.co.uk
Bike Trip, BikeTrip.org
Boneshaker Magazine, BoneShakerMag.com
Busch + Muller, bumm.de. Bike lights.
Car Free, CarFree.com/link/fpor.html
Christiania Bikes, christianiabikes.com
Climate Cycle, Chicago, IL; ClimateCycle.org
CrimAnimalz, CrimAnimalz.com

Critical Mass, Critical-Mass.info

Culture Change, CultureChange.org

Cycling News, CyclingNews.com. Australian racing bike news

Cyclists Inciting Change Thru Life Exchange, CICLE.org

David Hembrow English delivery bike baskets, Hembrow.eu/delivery

The Dog Trike, oaktreevet.co.uk/trike/trike.htm

European Cyclists Federation, ecf.com

Flying Pigeon LA, 3714 N. Figueroa St., Los Angeles, CA 90042;
    flyingpigeon-la.com

Free Wheelin Farm, Santa Cruz, CA;
    freewheelinfarm.com/home.html. An organic farm run by bicyclists
    who even deliver produce to local restaurants by bike.

Henry Work Cycles, workcycles.com/workbike

International Federation of Bike Messenger Associations.
    messengers.org. Provides resources for bike messengers worldwide.
    Contains information about bike messenger services in many towns
    and cities through the world.

International Human Powered Vehicle Association, ihpva.org

Kinetics, kinetics.org.uk. Folding bikes.

League of American Bicyclists, BikeLeague.org

Midnight Ridazz, MidnightRidazz.com

Organic Engines, organicengines.com/

Pedal People, Northampton, MA, pedalpeople.com. A human-powered
    delivering and hauling service.

Rickshaw Forum, rickshawforum.com

Ride A Cycle, RideACycle.org

San Diego County Bicycling Coalition, San Diego, California;
    SDCBC.org

Shifting Gears, ShiftingGearsCycling.com. Santa Monica, CA. Bike
    club.

Stolen Bike Registry, stolenbicycleregistry.com

Ten Links, TenLinks.com/cycling/portals.htm

Velo News, VeloNews.com. Boulder, Colorado

VeloVision Magazine, VeloVision.co.uk

Workbike, zerocouriers.com/workbike. Great source of information
    about bike-based business, bike trailers, and bike messenger services.

Workbike mailing list, ihpva.org/mailman/listinfo/workbike

Wrench Science, WrenchScience.com. For bicycle parts

Information about
various aspects of the plant-based diet:

I wrote my other book *Plant-based Regenerative Nutrition* to help people understand the physical and environmental benefits of following a plant-based diet. The book was published in 2023.

"In the past half-century, most U.S. livestock production has moved from small family farms to factory farms – huge warehouses where animals are confined in crowded cages or pens, or in restrictive stalls. The competition to lower costs has led agri-business to treat animals as mere objects, rather than individuals who can suffer.

Hidden from public view, the cruelty that occurs on factory farms is easy to ignore. But more and more people are taking a look at how farmed animals are treated and deciding that it's too cruel to support."
– Vegan Outreach, VeganOutreach.Org

"Vegetarian diets offer a number of nutritional benefits, including lower levels of saturated fat, cholesterol, and animal protein, as well as higher levels of carbohydrates, fiber, magnesium, potassium, folate, and antioxidants such as vitamins C and E, and phytochemicals. Vegetarians have been reported to have lower body mass indexes than nonvegetarians, as well as lower rates of death from ischemic heart disease. Vegetarians also show lower blood cholesterol levels; lower blood pressure; and lower rates of hypertension, Type 2 diabetes, and prostate and colon cancer."
– *Journal of the American Dietetic Association*; Vol. 103, No. 6; June 2003

"Vegetarianism preserves life, health, peace, the ecology, creates a more equitable distribution of resources, helps to feed the hungry, encourages nonviolence, and is a powerful aid for spiritual growth."
– Gabriel Cousens, M.D.

"All beings tremble before violence. All fear death, all love life. See yourself in others. Then whom can you hurt? What harm can you do?"
– Cautama Buddha

"Children who grow up getting their nutrition from plant foods rather than meats have a tremendous health advantage. They are less likely to develop weight problems, diabetes, high blood pressure, and some forms of cancer."
– Benjamin Spock, M.D.

"Eating meat leaves behind an environmental toll that generations to come will be forced to pay.
Land: Of all agricultural land in the U.S., 87 percent is used to raise animals. One acre of land can yield 20,000 pounds of potatoes or 165 pounds of beef.
Water: Raising animals for food consumes more than half of all the water used in the U.S. It takes 2,500 gallons of water to produce a pound of beef, but only 25 gallons to produce a pound of wheat.
Pollution: The meat industry causes more water pollution than any other industry. Animals raised for food produce about 43 tons of excrement every second. A typical pig factory farm generates raw waste equal to a city of 12,000 people.
There are ten times more pesticides used for animal feed than for food production that is used directly for human consumption. The pesticides, which don't readily degrade, enter our waterways and poison us, and our food.
Deforestation: Rainforests are being destroyed at a rate of 125,000 square miles per year to create space to raise animals for food. Fifty-five square feet of land are consumed for every quarter-pound fast-food burger made of rainforest beef."
– MeatStinks.Org

"I do not regard flesh food as necessary for us. I hold flesh food to be unsuited to our species. To my mind, the life of a lamb is no less precious than that of a human being. I should be unwilling to take the life of a lamb for the sake of the human body. The more helpless the creature, the more it is entitled to protection from humans, from the cruelty of humans."
– Mohandas K. Gandhi

"Animals and humans suffer and die alike. Violence causes the same pain, the same spilling of blood, the same stench of death, the same arrogant, cruel and brutal taking of life. We don't have to be a part of it."
– Dick Gregory

"The beef industry has contributed to more American deaths than all the wars of this century, all natural disasters, and all automobile accidents combined. If beef is your idea of 'real food for real people' you'd better live real close to a real good hospital."
– Neal Barnard, MD

"Pigs are basically the world's largest aquatic predator. It consumes more fish than all the world's sharks put together. The domestic housecat eats more tuna than all the world's seals put together. We're just pulling fish out of the ocean to convert it into fish meal for the raising of livestock, for pigs primarily. Sometimes cows, but generally, it's the pig and the chicken who are eating most of the seafood. And also for raising fish on fish farms. It takes 70 fish caught in the ocean to raise one salmon on a farm. It's an incredible waster of life in the oceans.

What I'm trying to do here is, there is a direct link. The environmentalists always say, "What has this got to do with us? This is animal rights." You cannot be an environmentalist unless you are a vegan or vegetarian. It's a total contradiction. It's willful ignorance on their part. You know, the Sierra Club or Greenpeace dismiss veganism or vegetarianism as an animal rights thing. It's an environmentalist thing. And in fact, it's probably the most significant environmental problem on the planet right now.

The meat industry releases more greenhouse gases than the entire transportation industry. The transportation industry releases CO_2 as a byproduct, which is a global warming gas. The meat industry not only produces CO_2, but also produces methane, which is a 26 times stronger greenhouse gas than CO_2. Plus nitrous oxide, which is 226 times as significant.

For instance, Greenpeace had an article like, "You can save the world by singing in the shower." Some sort of silly thing where you turn off the shower, sing, lather up, and the water you save when you turn off the shower will save the planet. And then you go and have a steak, which takes 3,000 gallons of water to produce one ounce. It's an incredible waste, and environmentalists don't take this into account.

Also, all the groundwater pollution in South Carolina is caused by the massive hog farms that are there. So the animal industry's causing groundwater pollution and wastes resources, and on and on and on and on. So it's definitely an environmental issue and environmental organizations don't want to confront this, in the same way that Al Gore doesn't want to confront this. Because they don't want to upset people."

  – Paul Watson of SeaShepherd.org, being interviewed by Doris Lin of About.com

"An act of violence against nature should be judged as severely as that against society or another person. The turning over of a stone, the unnecessary felling of a tree, or the slaughter of an animal is a crime to be weighed in judgment against the wants and needs of the person and the values of his society."

  – Michael J. Fox

"No member of the animal kingdom nurses past maturity. No member of the animal kingdom ever did a thing to me. It's why I don't eat red meat or white fish. Don't give me no blue cheese. We're all members of the animal kingdom. Leave your brothers and sisters in the sea."

  – Prince, rock musician

"I do feel that spiritual progress does demand, at some stage, that we should cease to kill our fellow creatures for the satisfaction of our bodily wants."

  – Mahatma Gandhi

"The world is not dangerous because of those who do harm, but because of those who look at it without doing anything."

  – Albert Einstein

"The meat industry spends hundreds of millions of dollars lying to the public about their product. But no amount of false propaganda can sanitize meat. The facts are absolutely clear: Eating meat is bad for human health, catastrophic for the environment, and a living nightmare for animals."

  – Chrissie Hynde

"There's no doubt in my mind that going vegetarian has made me feel better, not only physically but also because I learned about the suffering of animals who are raised and killed for food. I feel good knowing I'm not contributing to that."

  – Sophie Monk

"I stopped eating beef at 13 and stopped eating all meat a few years ago. I would feel guilty knowing that what was on my plate was walking

around yesterday. Either I could live with that, or stop eating meat. I choose the latter, and I'm happier for it."
– Carrie Underwood

"I don't eat any animals, or anything that has to do with animals. No fish, or egg, or dairy."
– Russell Simmons

"I've always felt that animals are the purest spirits in the world. They don't fake or hide their feelings, and they are the most loyal creatures on Earth. And somehow we humans think we're smarter – what a joke."
– Pink

"Nothing's changed my life more. I feel better about myself as a person, being conscious and responsible for my actions, and I lost weight, and my skin cleared up, and I got bright eyes, and I just became stronger, and healthier, and happier. Can't think of anything better in the world to be but be vegan."
– Alicia Silverstone

"Truly man is the king of beasts, for his brutality exceeds theirs. We live by the death of others: we are burial places. I have from an early age abjured the use of meat, and the time will come when men such as I will look on the murder of animals as they now look on the murder of men."
– Leonardo da Vinci

"We stopped eating meat many years ago. During the course of a Sunday lunch we happened to look out of the kitchen window at our young lambs playing happily in the fields. Glancing down at our plates, we suddenly realized we were eating the leg of an animal who had until recently been playing in a field herself. We looked at each other and said: 'Wait a minute, we love these sheep, they're such gentle creatures. So why are we eating them?' It was the last time we ever did."
– Paul and Linda McCartney

"Intellectually, human beings and animals may be different, but it's pretty obvious animals have a rich emotional life and that they feel joy and pain. It's easy to forget the connection between a hamburger and the cow it came from. But I forced myself to acknowledge the fact that every time I ate a hamburger, a cow had ceased to breathe and moo and walk around.

If you don't want to be beaten, imprisoned, mutilated, killed or tortured, then you shouldn't condone such behavior towards anyone, be they human or not.

Basically we should stop doing those things that are destructive to the environment, other creatures, and ourselves, and figure out new ways of existing."
– Moby

"As soon as I realized that I didn't need meat to survive, or to be in good health, I began to see how forlorn it all is. If only we had a different mentality about the drama of the cowboy and the range, and all the rest of it. It's a very romantic notion, an entrenched part of American culture, but I've seen, for example, pigs waiting to be slaughtered, and their hysteria and panic was something I shall never forget."
– Cloris Leachman

"I do not like eating meat because I have seen lambs and pigs killed. I saw and felt their pain. They felt the approaching death. I could not bear it. I cried like a child. I ran up a hill and could not breathe. I felt that I was choking. I felt the death of the lamb."
– Vaslav Nijinsky

"We pray on Sundays that we may have light, to guide our footsteps on the path we tread. We are sick of war, we do not want to fight, and yet we gorge ourselves upon the dead."
– George Bernard Shaw

"You have just dined, and however scrupulously the slaughterhouse is concealed in the graceful distance of miles, there is complicity."
– Ralph Waldo Emerson

"Thousands of people who say they 'love' animals sit down once or twice a day to enjoy the flesh of creatures who have been utterly deprived of everything that could make their lives worth living and who endured the awful suffering and the terror of the abattoirs."
– Jane Goodall

"I have no doubt that it is a part of the destiny of the human race, in its gradual improvement, to leave off eating animals."
– Henry David Thoreau

"A man can live and be healthy without killing animals for food; therefore, if he eats meat, he participates in taking animal life merely for the sake of his appetite. And to act so is immoral."
– Leo Tolstoy

"So I am living without fats, without meat, without fish, but am feeling quite well this way. It always seems to me that man was not born to be a carnivore.

Our task must be to free ourselves by widening our circle of compassion to embrace all living creatures, and the whole of nature, and its beauty."
– Albert Einstein

"It's best to avoid eating meat out of compassion. Before eating the meat, think of where it came from, through cutting an animal's neck, against its will, and how much suffering the animal experienced. After thinking about that, you can't eat the meat.

Meat may be nice for the person eating it, but not for the animal who suffered so much and didn't die naturally. You can say prayers for the animal that was killed, but if you eat the meat you are still playing a small part in the death of the animal.

If everyone stopped eating meat then no more animals would be killed for that purpose."
– Lama Zopa Rinpoche

"What is it that should trace the insuperable line? The question is not, can they reason? Nor, can they talk? But, can they suffer?"
– Jeremy Bentham

"Chickens, pigs, and other animals are interesting individuals with personalities and intelligence. What people need to understand is that if they're eating animals, they are promoting cruelty to animals."
– Pamela Anderson

"I've been a vegan since I was about 3 years old, and involved in animal rights for years. I've seen a number of animal rights films throughout the years, none has affected me as profoundly as *Earthlings*."
– Joaquin Phoenix http://www.earthlings.com/earthlings/video-full.php

For more about the plant-based diet, see my book *Plant-based Regenerative Nutrition*.

Plant Trees!

Without trees, you wouldn't be alive

Trees and forests are one of the many keys to life on this planet. Without healthy forests, and large numbers of a variety of trees (and the kelp forests, and other forms of marine plants of the rivers, ponds, lakes, and seas), we would all die.

The average human living in North America uses paper and wood products that amount to hundreds of trees during their lifetime. To replenish the trees we use – and to plant even more than we use – is a noble venture.

Please be involved with growing, planting, and protecting trees and forests. Support organizations working to do the same.

At least 17,510 tree species are threatened

"In a 2021 report, they announced they had found 58,497 tree species, of which 17,510 were threatened. Since then, almost 2,800 of those have been labeled as critically endangered. Some 142 species are thought to be extinct in the wild."
   – Aislling Irwin, *The loneliest trees: Can science save these threatened species from extinction?*; Nature.com

"The most endangered tree in the contiguous United States is most likely a battered old oak hidden deep in a Texas mountain range. It's trunk is scarred by a wildfire. Its limbs are weak from a fungal infection. Its habitat is imperiled by climate change. Scientists only realized the species still existed after stumbling upon the ailing specimen during an expedition this spring. And without swift action, researchers warn, Quercus tardifolia could truly disappear.

The species is among some 100 U.S. trees staring down the barrel of extinction, according to a sweeping new assessment published Tuesday in the journal *Plants People Planet*.

Amid an onslaught of invasive insects, a surge in deadly diseases, and the all-encompassing peril of climate change, as many as 1 in 6 (of the 881 species of) trees native to the Lower 48 [U.S. states] are in danger of being wiped out, scientists say. The list includes soaring coastal redwoods, capacious American chestnuts, elegant black ash, and gnarled whitebark pine. Yet only eight tree species are federally recognized as endangered or threatened.

Until several years ago, scientists didn't even know how many tree species existed (the correct number is 58,497)."
  – Sarah Kaplan, *As many as one in six U.S. tree species is threatened with extinction*; The Washington Post; Aug. 22

"It's easy to feel that gloom and doom… the scope of the crisis is really, really great now. We're losing species before they even get described.

It's this big swath of life that's totally unstudied or understood."
  – Murphy Westwood, Morton Arboretum, Illinois

Learn of the trees native to your region, and plant some

Learn the types of trees native to where you live. Some of the rare and endangered trees are native to your region. Find out which ones they may be, and then plant and protect them.

"Trees shade our ground, create topsoil, clean the air, and help the land attract, hold, and filter water. The trees and their roots purify the water as the rains fall. Clean streams keep millions of aquatic and other species alive."
  – Tim Hermach, President Native Forest Voice; *Forest Voice*; ForestCouncil.org; Spring 2006

"Destruction of forests is a leading cause of global environmental breakdown, including global warming."
  – AncientTrees.org, 2006

"We have nothing to fear – and a great deal to learn – from trees, that vigorous and pacific tribe which without stint produces strengthening essences for us, soothing balms, and in whose gracious company we spend so many cool, silent, and intimate hours."
  – Marcel Proust

American Chestnut Foundation, ACF.org
Alliance for Community Trees, ACTrees.org
American Forests, AmericanForests.org
Ancient Trees, AncientTrees.org
Budongo Forest, Budongo.org. African site.
Campaign for Old Growth, ancienttrees.org
California Rare Fruit Growers, CRFG.org
California Tropical Fruit Trees, TropicalFruitTrees.com.
Common Vision, CommonVision.org.
Earth First!, EarthFirst.org
Forest Advocate, ForestAdvocate.org
Forest Council, ForestCouncil.org
Forest Ethics, ForestEthics.org
Forests Forever, ForestsForever.org
Forest Protection Portal, Forests.org
Friends of The Trees, friendsofthetrees.net
Friends of the Urban Forest, FUF.net
Global ReLeaf, GlobalReLeaf.org
Gifford Pinchot Task Force, GPTaskForce.org
Living Tree Paper Company, LivingTreePaper.com
Australia's Men of the Trees, menofthetrees.com.au
North American Native Plant Society, NANPS.org
Native Forests, NativeForest.org
Natural Resources Defense Council, NRDC.org
Northern Nut Growers Association, NutGrowing.org
Protect Our Woodland, ProtectOurWoodland.co.uk
Rainforest Action Network, RAN.org
Redwood Rising, SaveTheRedwoods.org
World Rainforest Information Portal, RainForestWeb.org
Sanctuary Forest, SanctuaryForest.org
Save the Memorial Oaks Grove, SaveOaks.com
Save the Redwoods League, SaveTheRedwoods.org
Sequoia Forest Keeper, SequoiaForestKeeper.org
TasForests, TasForests.green.net.au. Tasmania, Australia.
Tree Musketeers, TreeMusketeers.org
Tree People, TreePeople.org
Tree For Life, TreesForLife.org
Trees for the Future, TreesFTF.org
Trees Foundation, TreesFoundation.org
We Save Trees, WeSaveTrees.org

"History will not only judge us by our mistakes, but by what we do to fix them."
– Jacques Yves Cousteau

"We never know the worth of water till the well is dry."
– Thomas Fuller, *Gnomologia*, 1732

"There is hope if people will begin to awaken that spiritual part of themselves, that heartfelt knowledge that we are caretakers of this planet."
– Brooke Medicine Eagle

"We need to realize that up until this point we have saved our own species with technology, new developments in agriculture, opening new land – and therefore of course destroying large numbers of other species. We've always found a way around our exponential population growth through technology. When it comes to energy extraction, we've had to develop very high technology and complex systems, and they're getting more complex all the time. We've reached a point where what we have wrought is so complicated and ill planned that we can't handle a lot of it. That takes us to a point where we have to recognize that we're not going to have any kind of livable planet for ourselves, unless we make our environment sustainable – and that includes the living environment. We have to slam on the brakes before we wreck the planet. Which we're about to do."
– E.O. Wilson, founder of the inquiry of sociobiology, and co-author of *The Theory of Island Biogeography*, and Pulitzer-prize winning co-author of *The Ants*; interview in OnEarth, winter 2011; OnEarth.org

Get involved in your local environmental and wildlife organizations. Get off fossil fuels. Replant paradise. Rewild Earth. Unpave the landscape. Restore wildlife habitat. (Book: *Ever Green: Saving Big Forests to Save the Planet*, by John Reid and Thomas Lovejoy.)

BeyondFossilFuels.org
BiologicalDiversity.org
Defenders.org
EarthEcho.org
EarthFirst.org
EarthIsland.org
EarthJustice.org
NoFracking.com
NRDC.org
ReWild.org
SaveTheRedwoods.org
SeaShepherd.org

Daniel John Carey

Humanity Needs An Intervention

Start the conversation

I wrote this book because I see what is happening to Nature, and because I think there are conversations humans need to have, things they need to consider, and actions they need to take. That is, if they are interested in not ending both humanity and life on Earth – this century.

Humans need an intervention for how they are misusing and abusing Earth, and the plants and non-human animals on her.

I attempted to make a list of the worst monsters now damaging the planet – that is, beside the most obvious: humans. What a coincidence that the list is all of things relating to human activity. The list includes: animal farming, burning fossil fuels, fracking, cattle breeding and grazing, climate change, dead zones, deforestation, desertification, factory farming pollution, human population explosion, killing wildlife, landscape chemicals, litter, mansions, medical pollution, military pollution, monocropped fields of animal feed, mountaintop removal, ocean acidification, offshore drilling, oil palm plantations, pesticides, pharmaceutical pollution, plastic pollution, poaching of wildlife, pollinator population plunge, private jets, shoreline destruction, tar sands mining, trawling, trophy hunting, urban sprawl, wildlife habitat loss, and so many other things that can't all be covered in a single book.

The planet is experiencing stress and destruction on every level.

Many species of plants and animals have already been sent into extinction. Many more are on the edge of extinction, and pretty much all are on their way there. All because of the most destructive creatures ever to exist on the planet: humans.

There is too much going on with the environment and wildlife to keep up with. I could read all day every day and not learn of it all. Reports from all over the world continually are detailing the loss of wildlife, the disruption of the environment, climate change, and the plight of humanity.

As I'm writing this paragraph in December 2023, scientists from all over the world are saying humanity is unlikely to survive past the later decades of this century.

Humans need a broad variety of plants and animals to survive. Including to keep the soil, water, and air, and all of Earth healthy.

The increased heat is making the plants so hot that photosynthesis is interrupted. That means, they die. This means weather gets interrupted, oxygen levels plunge, greenhouse gas sequestration is so reduced it only makes the land, rivers, lakes, marshland, and oceans even hotter. If there is a collapse, air conditioning and refrigeration systems will no longer work, and we will also be stuck in heat.

The number of species lost because of the climate change happening this century is one of the many things that will make it impossible for humans to survive.

The edible plants will be gone.

Pollinating species will be non-existent.

Scientists say that water from melting ice will take over coastlines and islands where hundreds of millions of people currently live. Where will they go? What will they eat? How will they survive?

Clouds will gather so much vapor the storms and floods will be incredible, ruinous, and alter landscapes in ways never experienced by humans. This will make it impossible to maintain farmland.

The desertification of land from dying plant and microorganism life will be an unending problem. Dust storms will become more common, because there will be no more roots to hold things in place, and no plants, bacteria, fungi, and wildlife to function in ways that make new soil.

There will be none of the basic functions of society. There will be no systems providing electricity, water, food, or fuel, and road, railroad, airport, or sewage system maintenance.

There will be no functioning transportation of any kind, including cars, trucks, busses, motorcycles, airplanes, ships, or trains. Fuel and/or electricity needed to run them eventually will not be available. Maintenance, repairs, and parts will become unavailable.

Schools and other government services will end.

Prisons and jails will have no security systems.

Police departments and the military will not function.

There will be no money worth anything.

Gold and silver will become worthless.

Real estate values will collapse.

There will be no imports or exports.

Technology will become useless.

There will be no phone service, TV, radio, or working satellites.

Sewage systems will become inoperable.

There will be no emergency services.

There will be no way to know what is happening beyond you region.

You will lose contact with anyone beyond your immediate vicinity.

Foods outside your region will become unavailable to you. Local food will become more and more limited.

Climate change is already impacting everyone on the planet.

It is 2023, and some foods are less available. Food prices are rising.

The Northern Hemisphere had the hottest summer on record.

As I write this, summer has started in the Southern Hemisphere. Temperatures are already breaking records from South America to Africa to Australia. Oddly, one way people learned about the heat wave in South America is because a young girl died of dehydration at a stadium concert of a globally famous pop singer.

More land and water animals are dying because of the heat, floods, storms, and loss of habitat.

More farmland is being ruined from drought, floods, pollution, road building, and urban spawl.

More pollution continues to be spewed into the atmosphere, and across the land, and is being absorbed by the water.

Ocean acidification is increasing.

Wildlife food sources are being reduced.

Trawling ships continue killing millions of sea creatures every day.

Billions of farmed animals are being fed with crops grown on monocropped fields stripped of native plants, cleared of native wildlife, and sprayed with toxic chemicals that spread into the air and water.

Millions of wild animals will have been killed again this year across the planet by farmers protecting their massively overbred cows, sheep, goats, chickens, pigs, and other farm animals, and to protect feed crops grown as food for farmed animals. Millions of dollars of government money is spent to support these killing of wildlife.

Massive amounts of plastics continue to be made and end up as trash across the land, and in rivers, lakes, marshes, swamps, and oceans. The plastic then kills wildlife. Nobody knows how long the plastic pollution will be around, as it will continue to kill more wildlife. Plastic pollution will be here longer than humans.

Pharmaceutical drugs, landscape and farming chemicals, and industrial pollutants continue to bioaccumulate in tissues of wildlife, resulting in lower birth weights, birth defects, still births, health problems, and more creatures on the edge of extinction.

What can be done?

Right now, humans need to discuss these things.

Massive actions need to be taken, on each continent, and on populated islands. The actions include planting billions upon billions of native trees and other plants, and improving conditions of wildlife and its habitat on every level.

It needs to be done quickly.

What you can do is be a part of the solution.

Start now.

One thing that can help Nature is to start using the most useful plant on the planet: industrial hemp. An acre of it can absorb more greenhouse gasses than an acre of trees. Hemp can be used instead of trees to make compostable paper, bags, cardboard, plywood, flooring, and packaging. Hemp can be used for hempcrete, instead of concrete. Hemp can provide food from the seeds. It can be used to make compostable plastics. It can be used to make ink, soap, lotions, and fabric. The resin and fiber can be used to make fiberglass. The cellulose can be used for insulation, and for making cellulosic ethanol that runs gas engines, and the seed oil can be used for diesel engines.

Hemp can be grown in every region. Many people can be involved in growing it and making products from it.

Research and learn about the uses of hemp.

A thing that will be of great help in protecting the climate, the air, water, and land, and wildlife is for humans to switch to plant-based diets. Fields currently used to grow feed crops and to graze cattle can be brought back to landscapes of native plants where wildlife can thrive. This will also reduce ailments related to eating meat, dairy, and eggs – including cardiovascular disease, arthritis, diabetes, kidney disease, and other chronic and degenerative health issues. This will result in fewer pharmaceutical drugs being used, less medical pollution, and healthier communities.

Another thing you can do is to replant what was the paradise of your region. This includes planting native trees, native brush, and native flowering plants to support wildlife, to build healthier soil, to protect water, to sequester greenhouse gasses, and to provide food for both wildlife and humans.

You and the people around you can all get involved in restoring native landscapes in your region.

Be a person actively involved in causing solutions.

Nuclear arms

All of us could be impacted – and most dead within hours, days, and weeks – of a nuclear war breaking out. The threat of this happening needs to be de-escalated. It has only been magnifying.

As Jimmy Carter said in his 1981 farewell address:

"It's now been thirty-five years since the first atomic bomb fell on Hiroshima. The great majority of the world's people cannot remember a time when the nuclear shadow did not hang over the Earth. Our minds have adjusted to it, as after a time our eyes adjust to the dark. Yet the risk of a nuclear conflagration has not lessened. It has not happened yet, thank God. But that can give us little comfort, for it only has to happen once. The danger is becoming greater. As the arsenals of the superpowers grow in size and sophistication, and as other governments – perhaps even in the future, dozens of governments – acquire these weapons it may only be a matter of time before madness, desperation, greed, or miscalculation lets loose this terrible force. In an all-out nuclear war, more destructive power than in all of World War II would be unleashed every second during the long afternoon it would take for all the missiles and bombs to fall. A World War II every second – more people killed the first few hours than in all the wars of history put together. The survivors, if any, would live in despair amid the poisoned ruins of a civilization that had committed suicide."

(Search: Eliminating nuclear arms.)

How do we prevent the meltdown of hundreds of nuclear power plants around the planet, if electricity becomes unavailable?

We Must

Transition to a sustainable humanity

"Driving all this is the fossil fuel industry. Enabling it are political leaders unwilling to bring this industry under control and who promote policies such as offsetting and massive gas expansion that simply enable this industry to continue."
– Bill Hare, physicist and climate scientist, and CEO of Climate Analytics

We must get off of fossil fuels.

We must stop polluting the land and water with plastic.

We must drastically reduce meat consumption, and stop breeding and feeding billions of animals for meat.

We must stop using millions upon millions of acres of land to grow feed crops for farmed animals. We must return that land to the wild.

We must stop killing wildlife to protect livestock and feed crops.

We must ban chemicals harming bird and insect populations.

We must stop paving over more and more land, but unpave where possible. Earth needs to breathe, and soil organisms need to flourish.

We must restore wildlife habitat, including by planting hundreds of billions of trees, and by restoring creeks, rivers, lakes, wetlands, grasslands, marshes, and headwaters forests.

We must restore the fragmented rainforests.

We must stop destroying ocean life.

We must do what it takes to stop the permafrost, glaciers, and polar ice caps from melting.

We must get back to growing food on land with many varieties of edible plants, and rely more on locally-grown foods. And do so while practicing composting and organic growing methods.

We must reduce the use of pharmaceutical chemicals, including by eating more healthfully, and eliminating junk foods, toxic foods, and foods prone to increase the risk of chronic and degenerative diseases. This means, following more of a plant-based diet.

"People don't realize that toxicity is, I believe, moving faster than climate change, and will be more deadly.
What we have done is we have created a world which is hostile to life, in almost every form – including our own."
– Jeremy Grantham, 2023

We must transition to making products that are not so toxic, that are compostable, that are not going to cause problems for generations of wild animals, and that will not pollute the oceans, lakes, rivers, marshes, and other water bodies of Earth.

Industrial hemp is an easily-grown plant that can flourish on every continent, and absorb massive quantities of greenhouse gasses, while also improving the soil. We can start utilizing industrial hemp to the advantage of us, the land, the climate, and wildlife.

(Search: Uses for industrial hemp.)

Try harder

"I don't know how many more warnings the world needs. It's as if the human race has received a terminal medical diagnosis and there is a cure, but has consciously decided not to save itself.
But those of us who understand, and who care, just have to keep trying – after all, what other choice do we have?"
– Professor Lesley Hughes, board member of the Climate Change Authority and emeritus professor at Macquarie University

Use your power of one – which is you – to help Nature and Earth.
Look for, and engage in solutions.
Learn about seed saving. Gather the seeds from your local edible plants, and help spread them to where they will grow.
Learn about permaculture, regenerative agriculture, and organic home gardening.
Have a pollinator garden. Even if it is one pot of organically-grown flowering plants.
Here at the end of 2023, an article being passed around between environmental activists and scientists I know is this one on Nature.com: *Degrowth can work – here's how science can help.*
In addition to degrowth, look into "ecological economics."
Be part of the solution.

Glossary of Names

1

Index

A

D

E

H

I

J

L

M

T

About the Author

He has been a ghost co-writer on books by other authors.

He wrote the books *Dream Another Dream, Dream Your World*, and *Plant-based Regenerative Nutrition.*

He runs a screenwriting workshop in Los Angeles, and polishes screenplays for writers, directors, and producers. His book about screenwriting, *Screenwriting Tribe*, has been used as a text in film schools.

To book the author for podcasts, interviews, other media, panels, and events, email: ScheduleTheAuthor@gmail.com

Join the Facebook page for the book: Humanity: The Final Century.

"Perhaps the most fundamental and important truth humanity
– in all of its arrogance and self-righteousness – cannot learn is that
while humanity cannot live without Nature, Nature can live
without humanity."
  – Dr. Steve Best

End the Illusion

A division between life forms on Earth is an illusion. We and fungi breathe in what the plants put forth. Plants take in what we and fungi breathe out. Water and gases continually enter us through our skin, mouth, and nose, and exit our body through our skin, lungs, and eliminative organs. Water and gases continually flow in and out of plants and other life forms. Plants build soil, and nurture animals with nutrients. The defecation and decomposition of billions of animals builds soil, and nurtures plants. The electric fields traveling through the earth and the atmosphere also travel through us, plants, and the substances in and around us. Trillions of solar electron neutrinos always forming in the center of the sun continually flow through space and through us, the plants and other life forms, and all that is around and beneath us. Meteorite dust that falls to Earth also ends up in wild animals, the plants, the land, and us. We are the plants, fungi, and bacteria, and they are us. There are substances in us that were in tropical rainforests, in ocean kelp forests, in salt marshes, in the dinosaurs, in ancient warriors, in slaves and kings, in birds that flew in the mountains, in butterflies that fluttered among wildflowers in distant lands, in turtles that swam across oceans, and in fish that dwelled at the bottoms of the seas. We and all life forms on this planet are soil, rock, plants, animals, minerals, air, water, carbon, electromagnetic fields, meteorites, other space matter, and solar electron neutrinos.

Get busy helping to save wildlife and Nature.

BeyondFossilFuels.org
BiologicalDiversity.org
ClimateDefiance.org
ClimateUncensored.com
Defenders.org
EarthEcho.org
EarthFirstJournal.news
EarthIsland.org
EarthJustice.org
NoFracking.com
NRDC.org
RAN.org
Rebellion.global
ReWild.org
SeaShepherd.org
ThirdAct.org

Please share this book.

Give a copy to your local library.

Write a review of the book on Amazon,
on other sites, and on social media.

www.ingramcontent.com/pod-product-compliance
Lightning Source LLC
Chambersburg PA
CBHW050502210326
41521CB00011B/2286